Welcome to the *EVERYTHING*®

These handy, accessible books give you all you need to tackle a difficult project, gain a new hobby, comprehend a fascinating topic, prepare for an exam, or even brush up on something you learned back in school but have since forgotten.

FACTS

Important sound bytes of information

You can read an *EVERYTHING*® book from cover to cover or just pick out the information you want from our four useful boxes: e-facts, e-ssentials, e-alerts, and e-questions. We literally give you everything you need to know on the subject, but throw in a lot of fun stuff along the way, too.

ESSENTIALS

Quick handy tips

We now have well over 100 *EVERYTHING*® books in print, spanning such wide-ranging topics as weddings, pregnancy, wine, learning guitar, one-pot cooking, managing people, and so much more. When you're done reading them all, you can finally say you know *EVERYTHING*®!

ALERT

Urgent warnings

QUESTIONS?

Solutions to common problems

Dear Reader,
Thanks for picking up *The Everything® Astronomy Book!* We hope you
enjoy reading it as much as we enjoyed writing it.
    Astronomy is a different sort of subject, one that can be enjoyed
equally by regular people in their own backyards and by astrophysicists
in mountaintop observatories. Although astronomy can sometimes seem
daunting to a beginner, just about anyone can look up at the night sky
and enjoy the stars. Even a little knowledge, such as learning the
constellations, where to find the planet Jupiter, or when to look for a
meteor shower, can increase your enjoyment immensely.
    Keep reading to find out how you can discover a new comet or
asteroid, or record observations of a meteor shower or variable star.
And don't worry, there won't be any math required, and there's no
exam at the end of the book.
    We've tried to make *The Everything® Astronomy Book* as accessible
as possible for future astronomers of all levels. Just make sure that your
mind is open to new experiences, including viewing stars and galaxies,
seeing back in time, and learning about the origin of the universe and
what our eventual destiny will be.
    Clear skies!

THE
EVERYTHING®
ASTRONOMY
BOOK

Discover the mysteries
of the universe

Cynthia Phillips, Ph.D., and Shana Priwer

Adams Media Corporation
Avon, Massachusetts

To our children Zoecyn and Elijah
Special thanks to Dr. Marshall Gilula for love and support, and to Raven for proofreading.

EDITORIAL
Publishing Director: Gary M. Krebs
Managing Editor: Kate McBride
Copy Chief: Laura MacLaughlin
Acquisitions Editor: Allison Carpenter Yoder
Development Editors: Michael Paydos
　　　　　　　　 Christel A. Shea
Production Editor: Khrysti Nazzaro

PRODUCTION
Production Director: Susan Beale
Production Manager: Michelle Roy Kelly
Series Designer: Daria Perreault
Cover Design: Paul Beatrice and Frank Rivera
Layout and Graphics: Brooke Camfield,
Colleen Cunningham, Rachael Eiben
Michelle Roy Kelly, Daria Perreault

An Everything® Series Book.
Everything® is a registered trademark of Adams Media Corporation.

Published by Adams Media Corporation
57 Littlefield Street, Avon, MA 02322 U.S.A.
www.adamsmedia.com

ISBN: 1-58062-723-4
Printed in the United States of America.

J　I　H　G　F　E　D　C　B　A

Library of Congress Cataloging-in-Publication Data
Phillips, Cynthia
The everything astronomy book : discover the mysteries of the universe
/ Cynthia Phillips and Shana Priwer.
p. cm. – (An everything series book)
Includes index.
ISBN 1-58062-723-4
1. Astronomy–Popular works. I. Priwer, Shana. II. Title.
III. Everything series.
QB44.3 .P48 2002
520–dc21
2002008439

This publication is designed to provide accurate and authoritative information with regard to the subject matter covered. It is sold with the understanding that the publisher is not engaged in rendering legal, accounting, or other professional advice. If legal advice or other expert assistance is required, the services of a competent professional person should be sought.

—From a *Declaration of Principles* jointly adopted by a Committee of the
American Bar Association and a Committee of Publishers and Associations

Illustrations by Barry Littmann.
Photographs in interior and insert reprinted with permission and courtesy of noted institutions.
Illustrations on pages 18, 23, and 26 courtesy of Dover Publications, Inc.

This book is available at quantity discounts for bulk purchases.
For information, call 1-800-872-5627.

Visit the entire Everything® series at everything.com

Contents

Introduction

Have you always wondered about the sky? Have you looked up at the multitude of stars and tried to find a constellation, but gotten lost when you tried to connect the dots? Do you know which bright objects in the sky are stars and which are planets? Have you wondered what a shooting star is? Have you ever worried that Earth might be hit by an asteroid?

Since ancient times, humans have looked up at the sky in wonder. Stars have always been a source of speculation, and detailed stories were invented by ancient cultures to explain the meanings of what we see in the cosmos. Fables concerning gods, mythological beasts, and other flights of fancy have all been inspired by the night sky. Stars were used to tell the seasons, determine when it was time to plant crops or hunt certain kinds of game, navigate from one place to another, and find the way back home. Dramatic cosmic events, such as an exploding supernova or a comet, would often be interpreted as a prophecy of a great historical moment or impending doom.

FIGURE 0-1:
A spiral galaxy from a distance

(refer to page 278 for more information)

Courtesy of NASA and The Hubble Heritage Team (STScI/AURA)

FIGURE 0-2:
The Globular Cluster M15

(refer to page 278 for more information)

Courtesy of NASA and The Hubble Heritage Team (STScI/AURA)

A good understanding of the nature of the stars did not come until the twentieth century. With the invention of sophisticated telescopes and ways of measuring the distances to celestial objects, astronomers realized that some objects in the sky were much farther away than others. An accurate understanding of the immense distances involved in astronomy was finally available.

Astronomers eventually came to understand that the stars in the sky are part of our own galaxy, as distinguished from nebulae, galaxies, and other objects that are far beyond the reaches of our own corner of the universe. The light from some stars in our galaxy takes many hundreds or thousands of years to reach us. Some of the other objects in the sky are, in fact, completely separate galaxies, which contain thousands or millions of stars of their own.

Astronomy is a unique science because although a true understanding requires years of study, anyone interested in looking up at the sky can enjoy its wonder. Amateur astronomy clubs exist all over the country and around the world. People can meet, help each other build or use telescopes and binoculars, and share in observing spectacular celestial objects.

Amateur astronomers, ones who do astronomy simply for the fun of it, can make significant contributions to the field. Amateurs routinely discover new comets, which can be named after them, and make important observations of celestial objects and events such as meteor showers. Astronomy is one of the few fields where amateurs are not only welcome, but also a necessary part of the larger community. The sky is an enormous place, and interested individuals from all over the world need to contribute their observations.

Some amateur astronomers have expensive and sophisticated equipment such as computer-controlled telescopes, deep-sky catalogs, and other observation and calculation devices. However, you can get involved in the excitement of astronomy simply by finding a dark place on a clear night and looking up.

Whether you already know the difference between Ursa Major and Ursa Minor, or you have trouble finding the Moon, this book is for you. You will learn how to find north without a compass, how to distinguish between different celestial bodies, and what would happen if you got pulled into a black hole. Common myths will be debunked—you will learn why the Big Dipper is not an actual constellation, and that a shooting star isn't really a star at all.

Our picture of the universe has changed greatly due to advances in observational techniques and tools, and it is likely to change further as even more exciting advances and discoveries are made. Who knows what the future will bring—perhaps telescopes on the Moon, people on Mars, and maybe one day travel to another star! We have presented the current state of research in this book, but since astronomy is an ever-changing field, it is quite possible that some of the ideas we present here may be re-examined or even disproved in the future. That's part of the excitement of science in general, especially in an evolving field such as astronomy.

CHAPTER 1
Ancient Eyes, Same Sky

Our ancestors used the sky in religion, science, philosophy, and countless other aspects of their lives. Ancient Egyptians observed the heavens for religious and civic purposes; Native Americans used medicine wheels to study and interpret the skies; Chinese astronomers were some of the first to study sunspots. Astrology itself emerged as a way to use astronomy to predict certain events and life patterns.

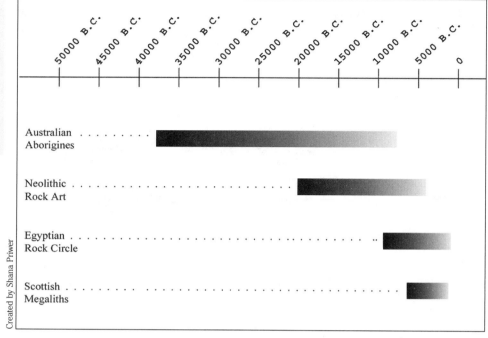

FIGURE 1-1:
Timeline of ancient astronomy 50,000 B.C. to 5000 B.C.

(refer to page 278 for more information)

Created by Shana Priwer

Archeoastronomy

Archeoastronomy is, as you might expect, a blend of archeology and astronomy. It combines the study of religion, folklore, celestial myths, and all ancient astronomical rituals and ideas. Often demonstrated through ancient writings and artifacts, archeoastronomy gives us a comprehensive, multicultural view of how modern astronomy came to be.

The origins of astronomical observation and representation date back at least as far as 20,000 B.C. While most Paleolithic art focuses on the representation of earthly creatures such as animals, Neolithic rock art with astronomical themes has been uncovered in Spain, Portugal, and France. These paintings and petroglyphs (rock carvings or drawings) depict the Sun, various planets, and several identifiable constellations. Sun images are often present in depictions of religious scenes, and much more about Sun gods is seen in later drawings and writings.

Prehistoric Astronomy

Stonehenge, located in England, is one of the more physical examples of an ancient culture celebrating the heavens. It consists of thirty stones, each approximately 30 feet in diameter. The entire structure was built in either four or five stages, probably between 3100 and 2300 B.C.

FIGURE 1-2:
Timeline of ancient astronomy 3000 B.C. to 2000 A.D.

(refer to page 278 for more information)

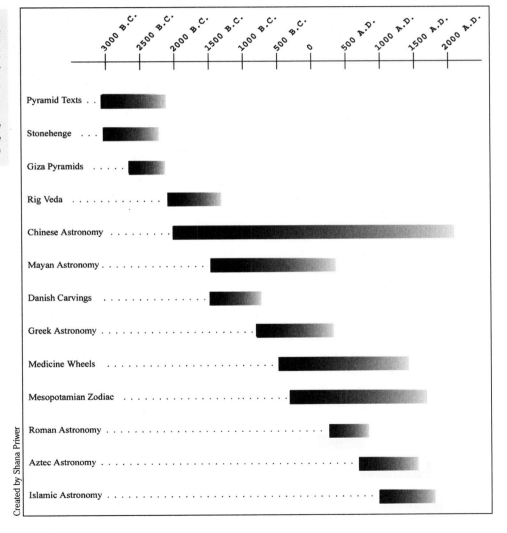

Created by Shana Priwer

FIGURE 1-3:
Stonehenge

(refer to page 278 for more information)

© 2001, www.arttoday.com

Stonehenge is oriented to align with the sunrise of the summer solstice; on that day, the Sun supposedly shone directly onto the Altar Stone. Stonehenge is generally considered to be a three-dimensional calendar. It is thought that Stonehenge was constructed to predict the optimum times for agricultural and religious events.

Astronomical observation was also an important part of Stone Age life in countries such as Ireland. Neolithic star stones dating back 4,000 to 5,000 years have been discovered, indicating an early awareness of celestial movements. The Solstice Stone found near Beltray marks both the summer and winter solstice; similarly, the Double Alignment stones at Barnaveddoge are aligned to both sunrise and sunset at the two solstices.

SSENTIALS

Stonehenge is located in Salisburg Plain, Wiltshire in southern England. Plan your visit carefully; on June 21 every year, from the Altar Stone, the Sun can be seen rising directly over the top of the Great Stone.

Astronomical monuments dating from early prehistoric times to the Bronze Age are also found in Scotland. Solstices were family and village events, and there is evidence to suggest that everyone in the village helped with the building process, from the quarrying of stones to the erecting of the rock circles. Located mainly in the northeast and southwest of Scotland, the circles are thought to mark a particular horizon point for different celestial events. Chambered cairns are also located all over Scotland. Cairns consist of stones that mark hidden chambers, and are thought to have served as vantage points for observing social or spiritual rituals.

Other countries and regions also have long astronomical histories. Australian Aborigines, some of the earliest astronomers, incorporated the heavens into their religious and civil lives at least 40,000 years ago. Aborigines are sometimes called the oldest astronomers in the history of time; the very word aborigine means "people who have been here since the beginning." Tool artifacts have been discovered from 20,000 years ago; artwork provides evidence to suggest that Aborigines used stars for the planning of rituals, hunting, food gathering, and navigation. Aborigines made accurate observations of celestial events, largely without the use of viewing instruments.

QUESTIONS?

Was there a real star of Bethlehem that the Magi followed, one that heralded the birth of Jesus?
Astronomers have suggested that the "star" could have been a supernova, a new comet, or the conjunction of two bright planets sometime between the years 8 B.C. and 4 B.C., which is when most historians think Jesus Christ was actually born.

The writings of Julius Caesar tell us that the Teutonic people, residents of Germany and Denmark, worshipped the Sun and Moon as far back as 1200 B.C. Rock carvings found in Denmark, dating to 1000 B.C., illustrate the Sun and fire as deities. A stone ring in the Danish city of Andebjerg clearly shows markings to observe sunrise at the summer and winter solstices and equinoxes.

What we can abstract from this study of prehistoric and early civilizations is that just about everyone, on every continent, figured out a way to observe the heavens and record their findings. Some countries attached extreme religious significance to the heavens, while others used the stars as a means to ensure subsistence. In all respects, we see the importance of astronomy in the religious, civil, and political lives of early peoples.

Egyptian Astronomy

Ancient Egyptians were extremely advanced in mathematics, astronomy, and physics. From 7000 to 6000 B.C., evidence of megalithic stones and stone rings suggests the beginning of early observations. A circle of standing stones located near Nabta bears a striking resemblance to Stonehenge, while predating it by over 1,000 years. The circle was likely used for the sighting and viewing of the solstices.

The Pyramid Texts, dating back to 3000 B.C., are the oldest surviving documents from the Egyptian heyday. Venus and Mars are referred to extensively, and the Pharaoh is constantly linked to the stars and planets. Star gods and goddesses played a critical role in the politics and religion of early Egyptian society. The star Sirius guided the creation of their calendar year, helping them predict when the Nile would flood and make the land fertile.

The three pyramids of Giza, arranged on a site called the Memphite Necropolis, were built between 2533 and 2505 B.C. In the Egyptian belief system, all Pharaohs were considered to be direct descendants of the Sun god Ra, so the connection to astronomy was a clear part of their everyday lives. Egyptians also believed that Egypt paralleled the cosmos, and was a direct representation of a land in the skies.

FIGURE 1-4:
Pyramid paralleling alignment.

(refer to page 278 for more information)

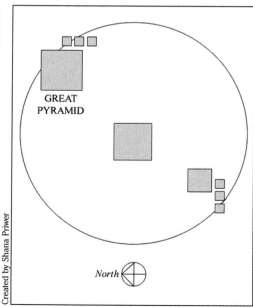

Designed as an eternal resting place for the Pharaoh, the Great Pyramids were constructed with careful planning. Stars were used to ensure that the building site faced exactly north. Each pyramid took more than twenty years to build. The Great Pyramid of Khufu is guarded by the enormous Sphinx monument. Built between 2589 to 2566 B.C., this pyramid contains an empty shaft from the queen's chamber. This shaft was originally thought to be for ventilation, but was later found to be directly aligned with Sirius.

The arrangement of these three pyramids has been interpreted to represent the three stars of the constellation Orion's belt. While the positions of the stars have changed over the last 4,500 years, it is possible to make adjustments for these changes and analyze the pyramids' positions based on what the Egyptians would have seen. Aerial views of the Memphite site reveal that the pyramids seem to correlate fairly well to the orientations of the Belt Stars.

Native American Astronomy

Astronomy was one of the basic components of Native American life. Seasons, spiritual ceremonies, time, and the calendar were defined by the stars. Three different aspects of the diverse Native North American cultures help us to understand the incorporation of celestial bodies into their lives: Chaco Canyon, medicine wheels, and information we have about the Lakota tribes. South American Native civilizations also used the stars to create detailed calendars, and for other purposes.

North American Cultures

Many different Indian tribes inhabited New Mexico's Chaco Canyon, including Aztecs in prehistoric times, then Anasazi and Navajo hundreds of years later. The canyon served as the homeland for a large farming society that used the stars to anticipate the changing of seasons, and therefore their livelihood. Buildings were planned to correlate directly with the alignment of the heavenly bodies, especially Sun, Moon, and Earth. Chaco residents constructed kivas, partially underground chambers used for spiritual rituals, which were aligned exactly with the summer or winter solstice.

Kivas may have been used for astronomical observation. Native Americans recorded supernovas and other astronomical events, possibly from these kivas. The supernova of A.D. 1054, which formed the Crab Nebula, was the best known of these events, and a petroglyph high on the canyon wall may depict this celestial event. Other artifacts also indicate their interest in the heavens, including the Sun Dagger, a rock formation found in Chaco Canyon, which may have functioned as an early calendar.

Medicine wheels, rock arrangements closely resembling spoked wheels, are found all over Native American territory. One of the earliest known medicine wheels was found at Moose Mountain in Saskatchewan, and dates to approximately 600 B.C. Many early medicine wheels were also created by the Plains Indians in regions all over the United States and Canada. The Big Horn Medicine Wheel in Wyoming, used between A.D. 1500 and 1700, is one of the best preserved of these wheels.

FACTS

Kivas were underground chambers constructed by several different Native American tribes. Spiritual rituals and astronomical observations—supernovas, comets, and even nebulae—might have been observed from kivas.

The Lakota Sioux Indians were particularly attuned to the movements of celestial bodies. They used the stars to guide them on hunting expeditions and to plan religious ceremonies. They used the direction and length of the Sun's shadow to determine when equinox was approaching, and based their calendar on these events. The Black Hills, a Lakota territory in South Dakota, contained several ceremonial locations. Each depicted a different constellation, as Lakota Indians considered the sky and earth to be mirror images of each other.

South American Cultures

The Mayans, a group of Native American civilizations living in different regions of Mexico, Guatemala, Honduras, and Belize, also used astronomy. Early Mayan culture began when civilization took root in Central America, around 10,000 B.C. The Pre-Classic Mayan period (1500 B.C.–A.D. 300) represented the cultivation of language, in addition to building and record keeping. The Mayans had a Sun god, but also studied several other solar system objects, including Venus and the Moon. Mayan rituals were coordinated with the stars, and the Mayan king created associations between himself, the cosmos, and life itself. Preserved Mayan stone tablet codices, such as the Codex Cortesianus and Codex Dresdensis, had hieroglyphs that represented various celestial bodies.

Aztec civilizations also made extensive use of the heavenly bodies in their religion and culture. Between A.D. 700 and 1400, Aztecs used myths of the gods to relate directly to the Sun and planets. One of their calendars was based on the solar year, indicating keen observation and record keeping concerning the Sun's movements. The motion of a star cluster, the Pleiades, aided the creation of the Aztec fifty-two–year cycle. The Calendar Stone, or Sunstone, is an Aztec artifact commemorating the Sun god and various Aztec periods.

Religious shrines and structures in Teotihuacan, Mexico, were laid out to align with the movement of the stars and other heavenly bodies. The main axis of the Pyramid of the Sun is said to line up directly with the setting of the Pleiades during the summer solstice.

Greek and Roman Astronomy

So much of where we are and what we know today is the result of ancient science and philosophy. Discoveries and observations recorded during the height of the Greek and Roman civilizations laid the foundation for the centuries of work that followed.

Greek Astronomy

Between approximately 700 B.C. and A.D. 300, the Greeks contributed greatly to astronomical observation. The representations of the cosmos achieved by the Greeks closely parallel the way we view the constellations today. Thales of Miletus was one of the first Greek mathematician-philosophers. As we know from the writings of Herodotus and Pliny, Thales was one of the first to accurately predict a solar eclipse through geometry and observation.

Pythagoras, most famous for the Pythagorean theorem, also made several critical discoveries in astronomy. After recognizing that Earth is round, he deduced that Venus as seen in the morning is the same star seen in the evening—an important step in understanding orbits. He was also one of the first to discover that the Moon's orbit is inclined to correspond with the equator of Earth. Euctemon and Meton, two other philosophers working in the fifth century B.C., created the first parapegma, a stone tablet that allowed people to make connections between dates and planetary movements. The parapegma was, perhaps, a very early predecessor to the planisphere (a map of the celestial sphere with a device that demonstrates which stars and constellations are visible for a particular date and location; see Chapter 12).

QUESTIONS?

If months are based on the lunar cycle, why aren't they the same length?
Recognizing that a strict thirty-day month got out of phase rather quickly, an Athens statesman named Solon made improvements by aligning the calendar with the phases of the Moon, alternating thirty-day and twenty-nine–day months.

Euxodus, a student of Plato, understood that Earth is but a small part of the larger cosmos. He also derived that Earth travels in an elliptical orbit around the Sun once a year, and turns about its own axis once a day. The same ideas were later developed by Aristarchus, an astronomer from Samos, who wrote and spoke extensively about the idea that all planets, Earth included, rotated around the Sun. However, this heliocentric theory was overshadowed by the more popular geocentric theories of the time, and fell into disregard until it was later discovered and proved during the Renaissance (Chapter 2).

Hipparchus, writing in the second century B.C., discovered the concept of the epicycle, the idea of bodies traveling in small circles that are part of larger circles. Epicycles were used to explain the motions of the planets as seen from Earth, and lent great support to the geocentric theory of the universe that placed Earth at the center with everything rotating around it. Aristotle was a great proponent of the geocentric theory, and his support led to its general adoption until the heliocentric, elliptical orbit theory was thoroughly understood many centuries later. Ptolemy, a Greek astronomer who resided in Egypt, consolidated and expanded upon the ideas of Hipparchus and Aristotle in his book *Almagest*. This work is our main reference for Greek and Roman astronomy.

Roman Astronomy

The heyday of the Roman Empire began with Caesar Augustus in 27 B.C., ending with Romulus Augustus in A.D. 476. While less is known about Romans and astronomy, we do know that they also made extensive use of astronomical observation. Denarii, Roman coins, depicted stars and planets but also recorded astronomical events such as a solar eclipse in 120 B.C. Can you imagine an eclipse on a United States coin? Probably not, which shows how important the heavens were to the Romans.

FACTS

The stars Mizar and Alcor were used by the Romans to test for vision. They appear together as one in the constellation Ursa Major, although in reality they are located many light-years apart. Mizar and Alcor are a double star, in this case an optical binary, because they're along the same line of sight from Earth.

Early Astronomy Elsewhere

While many people may be aware of Western civilization's contributions to early astronomy, it's worthwhile to note that Eastern cultures have been equally active in their observations of the stars and their significance.

Chinese Astronomy

Chinese astronomy is an ancient science and made advances much earlier than its counterpart in Western civilizations. The first solar eclipse recorded by the Chinese dates back to 2136 B.C. Early Chinese rulers generally had a court astronomer, whose job it was to record astronomical events for use in planning city functions. Artifacts show written records of comets and meteors as early as 1500 B.C.

By 300 B.C., Chinese records show a catalog identifying nearly 1,500 different stars, identified by their distance from the north celestial pole. Around this time, it is believed that the Chinese built intricate astronomical observation equipment such as the armillary, a spherical construction used to measure the longitude and latitude of planets and stars. In the second century A.D., they applied the mechanical waterwheel to the armillary to mechanize the turning of its metal arms. One of the most famous armillaries was built by a famous court astronomer named Guo Shoujing (A.D. 1231–1316).

The Chinese used other astronomical tools for observation and recording, including the quadrant (used to measure the altitude of celestial bodies), the theodolite (used to measure both altitude and azimuth) and the sextant (used to measure angular distance between planets and stars).

QUESTIONS?

What are altitude and azimuth?
Altitude is the number of degrees above the horizon an object is located. Azimuth is the direction of an object in the sky in terms of terrestrial compass directions (north, south, east, west). Together, altitude and azimuth provide coordinates for any object in the sky by combining the compass direction of the object with its height above the horizon.

By A.D. 30, they were studying the phenomenon of sunspots, which were not discovered in Western culture until much later. By around A.D. 200, Chinese astronomers were correctly predicting both solar and lunar eclipses. While much of Europe was stagnant during the Dark Ages, China made leaps and bounds in science and astronomy. A supernova explosion was witnessed and recorded in 1054. Magnetic compasses were in use by the 1100s. In 1439, Chinese ruler Zhengtong commissioned the Beijing Observatory, a large platform containing Qing Dynasty–era observing tools.

Japanese Astronomy

Stories and folklore in Japan often centered on astronomy during the Heian Era (A.D. 794–1185). A story about the stars Vega and Altair, passed down over time, led to the formation of the Tanabata Festival in Japan. Royalty were directly associated with the heavens, and it was believed that the death of a royal family member was connected to the birth of a star. The short version of this particular story is that the star Vega was actually the emperor Tentei's daughter Orihime, and Altair was her suitor Kengyuu; although they loved each other very much, the emperor allowed them to meet but once a year, and the celebration of this day became the festival of Tanabata.

Astronomy and folklore were also intimately linked in Japanese history. Japanese culture had a particular association with Orion. Rather than linking gods or mythology with this star grouping, Japanese tradition associates Orion with culture and public values. For example, the tsuzumi, a type of Japanese drum, is often seen as formed by the stars Betelgeuse, Gamma Orion, Rigel, and Kappa Orion. The symbolic usage of this drum carried down to other religious and even sporting events, such as the ceremony held before every Sumo wrestling tournament. Other tools or implements were contained in the belt of Orion, depending on the work of the region. This sort of customized interpretation seems unique in the study of the origins of astronomy.

Arabic Astronomy

Arabic astronomers made significant contributions to our knowledge and identification of cosmic elements. Basing much of their work on Greek developments, Islamic scholars read Ptolemy extensively and continued his work during Europe's Middle Ages. Much of the Islamic contribution to astronomy came from the strong ties it had to religion; the Islamic calendar was based exclusively on the Moon, and the holy month of Ramadan is based on the first appearance of the crescent Moon. Arabs were the first to be able to predict this sighting reliably. They used mathematics and instruments to derive precise measurements concerning the Sun, which contributed to their planning of daily prayer sessions.

Islamic observatories were built in Baghdad in the ninth century A.D. One of the largest observatories, honoring Ulugh Beg (1394–1449), was built in Samarkan in 1424. This circular building was home to a quadrant with a radius of more than 40 meters, indicating the importance of astronomy in their daily lives. Abu Abdullah Al-Battani (858–929), an Arabic astronomer and mathematician, worked extensively on calculating the Moon's orbit and the solar year. Many stars today still have Arabic names, such as Aldebaran (Ad-Dabaran), Altair (At-Ta'ir), and Deneb (Dhanab ad-Dajajah).

There is little in the way of biblical astronomy. Old Testament Jews made scarce mention of the planets and stars, except for Saturn and Venus. A reason for the Israelites' apparent lack of astronomical interest might be that Judaism, as a monotheistic religion, had little interest in focusing attention on anything in the heavens other than their one God.

Indian Astronomy

Astronomy in ancient India has a rich history dating back to its first mention in the Rig Veda around 2000 B.C. The Sun, stars, and bodies such as comets were actually deified in these early writings, indicating the crucial role the heavens played in people's religious lives. Astronomy and astrology were intermingled; the positions of the planets were a major part of both fortune-telling and religion.

One of the most famous Indian astronomers, Aryabhatta, lived and worked in the fifth century A.D. He reportedly studied astronomy at the University of Nalanda, and worked on pre-Copernican theories of heliocentric gravity. He recognized the similarity between the Sun and other stars, and the role the Sun played in gravitational force.

Some of the most important astronomical discoveries can be made without instrumentation. Ancient astronomers had no sophisticated viewing equipment, and were able to fully develop the art of naked-eye observation.

Most of Aryabhata's research and findings were written in the *Aryabhatiya,* which was translated into Latin in the thirteenth century. Many mathematical formulas were uncovered, such as equations for finding square roots and cube roots (although by that time they had been created elsewhere). Indian astronomers, working without telescopes, were able to predict eclipses with astounding precision, and they also used rough instruments to estimate the circumference of Earth.

CHAPTER 2

Astronomy As a Science

Renaissance astronomy forever changed the way we view our place in the solar system. This period in European history (from about the fourteenth to seventeenth centuries) has it all—discovery, fantasy, jealousy, and burning at the stake. Once new concepts were accepted, though, the path was paved for the development of modern astronomy as a science.

Pre-Renaissance Theory

Until the Renaissance period, the predominant view of the universe was geocentric—everything revolved around Earth. The Catholic Church was quite pleased with this model, originally presented by Ptolemy in the first century A.D., because it posited that the Sun and planets traveled in perfectly circular orbits around Earth. Placing Earth in the center of the universe seemed a divinely correct thing, and questioning this theory was like questioning God himself, a major taboo in a world where religion and justice were often one and the same.

Epicyclic Model

The old epicyclic model (planets traveling in small circles that are part of larger circles, as mentioned in Chapter 1) used to explain a geocentric universe was horribly complex, requiring advanced diagrams and theories that were never understood by most people. Ptolemy and his followers used epicycles to explain retrograde motion. While planets rotated about small axes called epicycles in this theory, they also orbited in larger circles around Earth; the larger orbit was called the deferent. When the epicycle and deferent coincided, the planet appeared to be moving in the opposite direction in a phenomenon known as *retrograde motion.*

Forcing the motions of the planets into the epicyclic model was a contrivance, in part designed to assure the Catholic Church that Earth and God were the center of the universe. Luckily, this theory was challenged during the Renaissance, opening the door for true scientific enlightenment.

The Church

During the early Renaissance, Church officials expanded the role of the Church from pure religion to secular events. The Church wanted to control the exploration of science, education, and many other secular pursuits. Such interests led to increased wealth and what many perceived to be corruption in the Church. Record numbers of clergy were having illegitimate children, and general morale was quite low. One of what some consider to be the Church's lowest moments in history was the sale of indulgences; for a price, people could buy their way out of purgatory and into heaven.

Martin Luther and his 95 Theses of 1517 presented the greatest challenge the Roman Catholic Church had ever seen in the form of the Protestant Reformation. Luther and his followers disapproved of the Church taking any interest in power and wealth. The creation of Lutheranism rocked the Church's foundation because it was no longer the one and only focus of Christianity. John Calvin (1509–1564) followed suit in France, further challenging the relationship between religious and political leaders.

ESSENTIALS

The relationship between the Church and science during the Renaissance was especially strained. With the Church's political power came control of money, information, laws, and public opinion. So strong was its hold that it took over 200 years for the advances in science and astronomy, among other things, to be disseminated and accepted.

There had been attempts to reform the Church prior to the Renaissance, of course. John Wycliffe (1330–1384) attacked the wealth of the Church in his Oxford University lectures, and John Huss (1372–1415) continued Wycliffe's ideas at the University of Prague. However, the controversies came to a head in the Renaissance, and these issues were the ones that the great astronomers of the period were up against.

The Renaissance Boom

The Renaissance brought a boom as new ideas and techniques flourished in art and architecture, as well as in astronomy. Many major religious structures and hospitals were designed and built during this period, which had a huge impact on art and architecture for future generations. Unlike astronomy, paintings and buildings stood as tangible, credible, and readily believable proof of the developments that occurred during this remarkable period.

Renaissance music also made its mark on history. Musicians used mathematical constructs to achieve certain types of chords and phrases. Intervals of thirds and multiple-line harmony were characteristic of the period. Johannes Ockeghem (1410–1497) introduced the concept of counterpoint, and the polyphony of the Italian madrigal took center stage.

The widespread interest in astronomy and science carried into music as well, particularly in the concept of music of the spheres. This idea was generated and took shape entirely during the Renaissance. Music of the spheres is the concept that the world is united by musical harmony; everything has its own music and rhythm, from people to the stars and planets. Revelations in astronomy, proven or otherwise, eventually made their way to many other areas of Renaissance life.

Copernicus

Nicolaus Copernicus (1473–1543) was one of the most significant players in Renaissance astronomy. A Polish-born physician, astronomer, philosopher,

Nicolaus Copernicus

and mathematician, he studied math and optics at Cracow University. Copernicus served as a canon at the Cathedral of Frauenburg for most of his adult life, and he gained tremendous influence through his many religious and political connections. Working before telescopes were developed for astronomers' use, he performed his observations almost exclusively with the naked eye.

Copernicus's greatest contribution to modern astronomy was the deduction of a heliocentric (Sun-centered) theory of the universe. He was one of the first astronomers to realize that planets orbit the Sun, not Earth. Copernicus replaced Earth with the Sun as the center of the universe, then deduced that the Moon orbited Earth. He did not require epicycles to explain retrograde motion; planets simply traveled at different rates as they orbited. While the epicyclic model was very complex, the heliocentric model of the universe was a simple and elegant diagram. Copernicus still used epicycles to explain the detailed aspects of planetary motions, partially because he incorrectly assumed all orbits to be circular instead of elliptical.

FIGURE 2-1:
Geocentrism
and
heliocentrism

(refer to page
278 for more
information)

Created by Shana Priwer

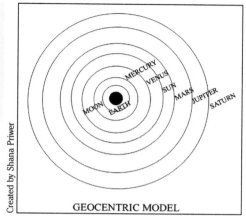

GEOCENTRIC MODEL

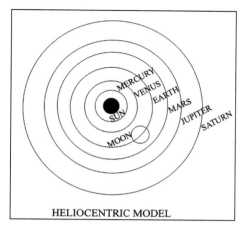

HELIOCENTRIC MODEL

Copernicus was also one of the first astronomers to understand that Earth rotates on its own axis daily, and orbits the Sun yearly. His ideas directly refuted the geocentric (Earth-centered) view of astronomy put forth by Ptolemy in the first century A.D. Copernicus's magnum opus of 1530, *De Revolutionibus Orbium Coelestium* (On the Revolutions of Celestial Orbs), was published postmortem in order to circumvent the religious fervor its publication would cause. Copernicus apparently feared not only retribution from the Church, but also mockery from other academics who dared not refute the conventional geocentric theory.

FACTS

Copernicus made his mark in history by understanding that Earth rotates about its own axis daily, and about the Sun yearly. Heliocentrism forever changed our model of the solar system.

Known as something of a perfectionist, Copernicus never felt his work was ready for publication, and might have been surprised at the impression it made on both the Church and academics. Fortunately, a German mathematician named Rheticus took interest in Copernicus's work in 1539, and sent it for publication after Copernicus died.

Tycho

Tycho Brahe (1546–1601) studied planetary motion precisely, and made extensive calculations concerning the positions of planets over time. His data would eventually provide much-needed support for Copernicus and the heliocentric theory of the universe.

Born into a family of nobles in Denmark, Tycho studied at various universities throughout Germany. He became interested in alchemy and astronomy, acquiring and mastering several astronomical instruments. He was a strong proponent of the use of instrumentation in obtaining astronomical data. Tycho is known as one of the best and most complete pretelescopic observers. In 1574 he discovered a new star (Stella Nova) in Cassiopeia, publishing a paper on it shortly thereafter.

FACTS

In 1575, King Frederick II of Denmark gave Tycho the island of Hven, where he built his Uraniborg Observatory. Tycho made painstaking measurements of the Sun, Moon, Earth, and other planets and stars. The precision of his instruments allowed him to make measurements that were unprecedented in accuracy.

Tycho formulated a model of the universe that was a compromise between heliocentrism and geocentrism. In his compromise theory, Tycho believed Earth was at the center of the universe, but that all other planets orbited the Sun instead of Earth. While he was unwilling to completely disregard the classical Greek notion of Earth as the center of the universe, he favored Copernicus's revised notion of epicycles that corresponded to the idea that other planets orbited the Sun. Tycho played a critical role in other areas of astronomy, and identified and cataloged over 1,000 stars, an unprecedented achievement.

After many years in Denmark, Tycho moved to Prague where he met a scientist named Johannes Kepler (1571–1630). As the story goes, Kepler's work impressed Tycho sufficiently that, upon his death, Tycho left his life's research to Kepler. Other versions suggest that Kepler obtained Tycho's data through somewhat illicit means.

Kepler

Kepler grew up in Germany, and his areas of study included theology, astronomy, and mathematics at the University of Tuebingen. He was a firm believer in Copernican heliocentrism, and analyzed the solar system in terms of concentric spheres. While his theories were ultimately incorrect, he pushed the envelope of current thinking about astronomy and opened the door for new theories and ideas.

SSENTIALS

Kepler and Tycho met in 1600, one year before Tycho's death. Kepler had been shut out of German academia in an attempt to rid the country of heretic influences. Ultimately, Tycho extended an invitation to come work with him in Prague, so Kepler studied with Tycho in his last remaining year.

Using Tycho's detailed and accurate measurements of planetary positions, Kepler was able to use his mathematical background to model the motions of planets better than ever before. Kepler's discoveries and assimilated information had enormous implications for physics and astronomy, and are known as Kepler's laws of planetary motion. Kepler was one of the first astronomers to understand that planets followed elliptical, rather than circular, orbits around the Sun, and this understanding eliminated the need for unwieldy epicycles.

Kepler's First Law

First, Kepler was able to determine that planetary orbits were elliptical in shape, with the Sun at one focus. All previous theories had put either the Sun or Earth at the center of a circular orbit.

Kepler's Second Law

Second, Kepler determined that planets moved faster as they travel closer to the Sun, rather than traveling at a constant speed as previously assumed. The point on a planet's orbit nearest the Sun is called *perihelion*, and the point at which a planet is farthest from the Sun is termed *aphelion*. Planets move their fastest at perihelion, and their slowest at aphelion.

Kepler's Third Law

Third, Kepler discovered that the time a planet took to complete an orbit was related to its distance from the Sun. This logic applies to our study of planets and orbits today. Pluto, the planet farthest from the Sun, has an orbit of 248 years; Mercury, the planet closest to the Sun, has an orbit of 88 days.

Kepler's Laws of Planetary Motion

First law: Planetary orbits are ellipses, with the Sun at one focus.

Second law: A line extended from the Sun to a planet will sweep equal areas in equal time.

Third law: The orbital period of a planet (squared) is proportional to its average distance from the Sun (cubed): $\dfrac{P_1^2}{P_2^2} \propto \dfrac{R_1^3}{R_2^3}$

Further Development of Kepler's Ideas

Giordano Bruno (1548–1600) took the ideas of Copernicus and Kepler a step further. Going directly against the philosophy of Aristotle and the astronomy of Ptolemy, Bruno suggested that space might be truly infinite, and that our Earth and Sun might belong to just one of any number of such systems. The Church saw these ideas as completely heretical and blasphemous; Bruno was arrested by the ecclesiastical Inquisition, and unfortunately condemned and burned at the stake in 1600.

Galileo

Adding to Kepler's work, major breakthroughs in Renaissance astronomy were made by Galileo Galilei (1564–1642). Born in Pisa, Galileo lived in a

Galileo Galilei

monastery until age fifteen. Later, he was expelled from medical school, became a mathematics tutor, and eventually a math instructor at the University of Pisa. His temper and high spirits were not conducive to keeping a job, and he transferred to the University of Padua in 1592. Galileo had a tumultuous personal life, and ended up tutoring the son of the grand duchess of Tuscany to help pay the bills.

Galileo's work went toward disproving Ptolemy's geocentric theory of the universe and supporting Copernicus's heliocentric model. Much of his early work was in physics; he developed one of the first theories in which gravity was distinct from an object's mass. The notion that he challenged Aristotle's theory by dropping objects from the Tower of Pisa, though, is probably just a myth.

Galileo's Use of Optics

In 1604 Galileo learned of the telescope through a Dutch optician. The telescope had been invented long before Galileo was born, and the model he built was actually based on an existing optical device in Holland. Although he had no plans or pictures of the Dutch device, he had sufficient, though scant, information to develop a significantly advanced model. Using sophisticated optics and a telescope he designed for looking toward the heavens, Galileo made multiple observations of specific celestial bodies. Because his discoveries became so famous, he is often credited with the invention of the telescope, but his research was actually based on years of development and experimentation by others.

QUESTIONS?

Why is Galileo often remembered for inventing the telescope?
Although Galileo is sometimes remembered for his invention of the telescope, he wasn't the first to create or build one! However, Galileo did develop a model that was much more effective for his purposes, and his significant observations became famous, which gave credence to the myth that he invented the telescope.

One of Galileo's most significant discoveries was the notion that there are millions of stars that cannot be seen with the naked eye. His advanced telescopic observations showed him a wide array of celestial bodies never before seen. He observed that the glow of the Milky Way came from clusters of stars, another phenomenon that had not yet been explained. His use of telescopic magnification taught him that while relatively close objects appeared magnified, stars that were very far away would not appear any larger through a telescope.

Galileo observed mountain ranges and craters on the Moon, as well as spots on the Sun; this information indicated that the planets were not perfect spheres as had previously been theorized. He realized that the planet Venus went through a full range of phases, much like the Moon.

In one of his most important observations, Galileo discovered four moons orbiting around the planet Jupiter. This discovery was extremely important; he proved that Earth wasn't the only planet around which other celestial bodies rotated. He published these findings in 1610.

Reactions to Galileo's Findings

Publishing of Galileo's discoveries produced varied results. Christopher Clavius, one of the leading Jesuit astronomers of the seventeenth century, used a telescope to confirm Galileo's findings with regard to the phases of Venus. While he was not ready to affirm Galileo's heliocentric theory, he did embrace Tycho's compromise. This act initially led to some success for Galileo, who made a trip to Rome in 1611. There he was given the support of Pope Paul V and was generally perceived favorably by the Catholic Church.

One of Galileo's most important findings was that Jupiter also had orbiting moons. For the first time, it was proved that Earth wasn't the only planet around which other celestial bodies rotated, and our place as center of the universe was permanently disturbed.

Through the 1630s, however, Galileo adamantly refuted the accuracy of Tycho's compromise theory and became a powerful advocate for Copernican heliocentrism. His strong personality prevented him from ceasing his campaign, and he insisted on trying to convert the Church, as well as the rest of Europe, to the Copernican model. His forcefulness called for the Catholic Church to reinterpret the Scripture based on what seemed to be an unproven theory. Part of the reason Galileo was subsequently discredited was because he insisted, despite proof from Kepler, that planets orbited the Sun in perfect circles; Jesuit astronomers knew that elliptical orbits made more physical sense, and thus were more suspicious of Galileo's other claims.

The Church's Role

Galileo's first trial was in 1616. Having recently survived the Protestant Reformation, the Catholic Church was very restrictive in its allowance for lenient interpretation of the Scripture. Galileo's findings irked religious leaders at all levels, and he was the recipient of a letter in 1615 from Cardinal Robert Bellarmine, who asked for further proof of his theories. Despite the conversations that ensued, Copernican heliocentrism was deemed heretical. Galileo was ordered to cease and desist in February 1616.

In 1633, after Galileo published his theories in *Dialogue on the Two Great World Systems,* there was an official Inquisition in Rome. He was condemned to house arrest for the remainder of his life, and was sentenced to recant publicly before his death in 1642. The Church banned his book until 1833, at which point heliocentrism was finally considered fact.

It was not until 1993 that Galileo was formally cleared of the charges of heresy. Pope John Paul II first ordered the Pontifical Academy of Sciences to study the case in 1979, and the report was presented to the pope in 1992. The general position of the Catholic Church was that, at that point, the pope was most interested in mending the rift which had developed between religion and science, in no small part due to Galileo's forced discrediting and arrest.

Newton

Sir Isaac Newton (1642–1727) is one of the best-known British alchemists, astronomers, theologians, and historians. He is the founding father of calculus, and he made significant advances with physics, optics, and light.

Sir Issac Newton

Raised by his grandmother, Newton was supposedly a poor student until he received a somewhat nasty bump on the head in grade school. This turnaround, be it fact or fiction, led to the creation of one of the most advanced scientific minds the world has ever known. Newton studied at Trinity College in England, and became a mathematics professor there in 1667.

Newton believed firmly in the idea that everything that happened had a cause. He was not, in this sense, a dualist in the tradition of Descartes; dualists believed that some items of truth could be taken without proof, such as the existence of God. Newton believed that all events, earthly and celestial, occurred for a reason.

In 1668, Newton built the world's first reflecting telescope. The history of the telescope does not begin with Newton (or even Galileo), but what makes reflecting telescopes unique is that light is gathered and reflected through a curved mirror, rather than a lens. All light is reflected equally, which eliminates the color and edge distortion found in refracting telescopes. Once light is gathered and focused on a point, a second mirror directs it to an eyepiece, where it can be magnified and viewed. Reflecting telescopes provided a major breakthrough in telescope technology; because the mirror can be structurally supported from behind (where a clear lens could not), reflecting telescopes can be built significantly larger than refracting, and the creation of the Newtonian telescope is one of Newton's most important legacies.

Newton is equally famous for his laws of motion, which explain many of the world's phenomena and are still taught to physics students worldwide. *Philosophiae Naturalis Principia Mathematica,* Newton's cornerstone physics treatise, showed how gravitation applied to both falling objects on Earth and celestial bodies in orbit. The basic idea of Newton's *Principia* is that objects in motion tend to stay in motion until an external force causes a change; the change is then proportional to the force that caused the change, and the object will go in the direction indicated by the change. These three ideas comprise Newton's Laws of Motion:

Newton's Laws of Motion

1. Objects in motion tend to stay in motion until an external force causes a change.
2. The change in motion of an object acted upon by an external force is then proportional to the force that caused the change.
3. The affected object will go in the direction indicated by the change.

In Newton's law of universal gravitation, he determined that the gravitational force between two bodies was proportional to the product of their masses, and that the gravitational force of a body on Earth was equal to its weight. While these ideas are common knowledge today, they were revolutionary at the time of Newton.

CHAPTER 3

How the Universe Came into Being

The Bible's Old Testament begins with the creation of the heavens and Earth from a great void. Thousands of years later, modern science has come up with a theory of creation that sounds remarkably similar. The study of the origin of the universe is called *cosmology*, and the formation of everything around us is a subject that touches people of all backgrounds and religions.

The Big Bang

In the beginning, say cosmologists, there was nothing. This beginning took place somewhere between 10 and 20 billion years ago. Scientists are still trying to pin down exactly how old the universe is (and you'll learn more about that later in the chapter).

Don't confuse *cosmology* with *cosmetology*! Cosmology is the study of the origin of the universe, while cosmetology is the cosmetic treatment of the hair, nails, and skin. If you walk into a beauty shop and ask to see a cosmologist, you'll get some strange looks!

To back up briefly, it isn't fair to say there was really nothing. There was all the *stuff* of the universe, that which makes up matter and energy and all the other physical aspects of our world. This matter was all compressed into a space smaller than the nucleus of an atom! The initial state of the universe is sometimes called a singularity, because that's all there was—everything compressed down to a single point. Space and time didn't even exist yet.

FIGURE 3-1:
A timeline of events after the Big Bang

(refer to page 278 for more information)

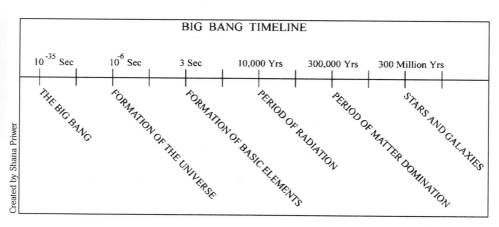

Then, suddenly, something happened. A huge explosion, too big to comprehend, sent all the universe-composing stuff flying out in all directions. This explosion was trillions of degrees in temperature, and initially the material strewn out was infinitely dense. For a tiny fraction of a second, the universe was an extremely dense, extremely hot fireball, and then the universe began to expand. This immense explosion created all the subatomic particles (quarks, photons, electrons, neutrinos, and others) that make up matter and energy. It also created time and space! This event was the point at which the clock of the universe started running; space began to be created as matter and energy streamed out in all directions. The event is what we call the Big Bang.

The Aftermath

After the giant explosion, as subatomic particles began to form, the universe was expanding at many times the speed of light. In less than a thousandth of a second, the universe expanded from the size of an atomic nucleus to 10^{35} meters in width (about 600 nonillion miles—or 600 with 30 zeros after it)! This event was called the *inflationary epoch*, and was almost perfectly smooth and symmetrical in all directions (isotropic). However, there were slight clumpings and inhomogeneous areas in the expansion, and it was these density fluctuations that eventually allowed the formation of stars and galaxies.

About one millionth of a second after the Big Bang, the inflationary epoch was over and the universe's expansion continued much more slowly. The universe was now a cloud of gas that was rapidly cooling and becoming less dense as it expanded. After about one second, the temperature had dropped to a mere 10 billion degrees. After three minutes, at a temperature of about 1 billion degrees, elements began to start forming. First nuclei of hydrogen, the simplest element, were formed, and then hydrogen nuclei began to combine in pairs to form helium nuclei. This process is called nucleosynthesis.

About 10,000 years after the Big Bang, the universe was in what's called the Radiation Era. Most of the energy of the universe was still in the form of radiation, mostly different wavelengths of light including x-rays, radio waves, and ultraviolet rays. This radiation marked the last remnant of the initial primordial fireball. As the universe continued to expand and cool, these radiation waves got stretched and spread out until they formed what's called the *cosmic microwave background radiation*, a quantity that can be measured throughout the universe.

FACTS

The name Big Bang was actually made up by a scientist who proposed a rival theory, in order to make fun of the Big Bang theory. Unfortunately for him, the Big Bang theory was ultimately proved right, and the disparaging name stuck.

After about 300,000 years, matter finally began to dominate the universe. At the 300,000-year mark, matter and energy were present in about equal proportions in the new universe. The universe was still expanding, though, and the leftover light waves kept getting stretched to longer wavelengths (and lower energies). Matter, however, does not get stretched out this way; it simply rides along on the wave of expansion. At this point in the history of the universe, the temperature was about 10,000 degrees Kelvin. Lithium atoms began to be formed at this stage, and electrons joined with the hydrogen and helium nuclei to make stable neutral atoms.

FIGURE 3-2:
Stars form in nebulas such as the Keyhole Nebula

(refer to page 278 for more information)

Courtesy of NASA and The Hubble Heritage Team (STScI/AURA)

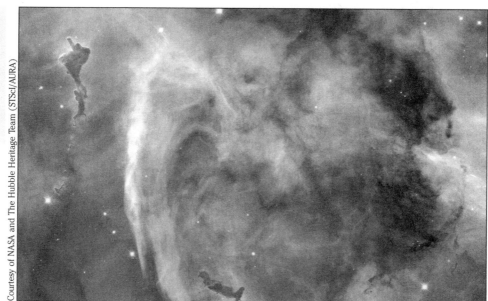

The universe continued along its path as a giant cloud of gas until about 300 million years after the Big Bang. By this point, slightly irregular areas of the gas cloud continued to attract more material, and these pockets of gas became denser and denser. Eventually individual stars collapsed and ignited in these pockets of gas, and clusters of stars formed the first galaxies. The period of the formation of the first galaxies was still 12 to 15 billion years before the formation of our own solar system. The universe has continued to form galaxies since then, and each galaxy is made up of millions of stars like our Sun. Each star in a galaxy can have its own planets orbiting it, thereby creating its own solar system. Our own Sun did not form until only about 5 billion years ago, and our solar system formed after that.

Creation Myths and Modern Science

Modern science, including the Big Bang theory, states that the universe came into existence with a giant explosion. Before this explosion there was nothing: no matter, no time, no space. This explosion created all the matter that currently exists in our universe, as well as the laws of physics that govern our lives. Throughout time, humans around the world have attempted to understand their origins. Where did we come from? Why are we here? Some of these ancient stories, handed down to us through writings or oral tradition, bear remarkable similarities to the story of creation that has now been proposed by modern science.

There are stories from cultures around the world of the creation of the universe out of a great void, and the creation of ordered material from a primordial, undifferentiated chaos. These themes parallel the Big Bang story of the universe coming into existence in an instant, along with the slow accretion of stars and planets from the primordial gas cloud. Do these connections have any great significance? Perhaps not, but perhaps these themes are basic in humanity's efforts to understand our origins. And perhaps the Big Bang theory is yet another story to be looked on as a creation myth by an even more advanced culture or civilization that could come after ours!

Scientific Evidence

As with any new theory, the Big Bang became more accepted as new discoveries and experiments continued to establish supporting evidence. The more researchers learn about what remains in the cosmos and how it relates to Earth and the solar system, the better they will be able to address remaining questions about our past—and our future.

Spectroscopy

So what did Hubble do? Well, in the early 1800s, scientists had discovered spectroscopy. Physicists working in this new field discovered that when any distinct chemical element was heated, it gave off light that had several dark lines in it, as if it were missing certain wavelengths (or colors) of light. Most objects put out light at a whole range of wavelengths just as a light bulb does. You can see these different colors by passing the light through a prism, which separates the light into individual colors. When the light given off by heating different chemical elements was studied this way, some of the colors were missing from the spectrum. Scientists were able to determine exactly which lines were missing from the spectrum of each particular element.

Scientists then used this knowledge to look at light from the Sun and other stars. By looking at what lines were missing from the light from stars, they were able to tell what elements were present in the Sun and in other far-off stars! This new information gave scientists exciting insight into the composition of stars, and proved that the same elements that made up the Sun were also present here on Earth.

The Doppler Effect and Redshift

Also in the 1800s, Christian Doppler discovered that when the source of a sound wave is moving, its pitch changes. If the sound is moving away from you, its pitch is lower, and if it's moving toward you, its pitch is higher. You notice this phenomenon when a fire truck or police car drives by you with its siren blaring. This concept also applies if the source of the sound is not moving, but the observer is. This effect (called the Doppler effect) also works for light. When an object is moving away from you, its light waves get stretched out and shifted toward the red end of the electromagnetic spectrum, where wavelengths are longer. If an object is moving toward you, its light gets squeezed up and its color moves toward the shorter, bluer end of the spectrum.

Astronomers were able to combine spectroscopy with Doppler shifts to study the motion of stars. The first star studied was the bright star Sirius, which has dark lines due to hydrogen in its spectrum. However when measurements were taken in the laboratory, these dark lines did not appear where expected on the spectrum; instead, they were shifted toward the red end. Scientists realized that this shift meant that the star Sirius was moving away from Earth! They could even measure the amount that the lines were shifted toward the red (called the redshift) and compute the speed at which Sirius was moving away from Earth.

Radial Velocity Versus Distance

Starting in the 1890s, astronomers at major observatories started measuring the radial velocities of various stars, as well as other objects then called gaseous and planetary nebulae. They began to notice that the velocities of nebulae, such as the Andromeda Nebula, were much higher than those of individual stars.

QUESTIONS?

What is radial velocity?
In astronomy, radial velocity is the line-of-sight difference between the velocities of Earth and the object moving away from (or toward) it. Redshift, which signifies an object is moving away, is one way to measure radial velocity.

Starting around 1913, astronomers including Edwin Hubble started to use special variable stars called *Cepheids*. The brightness of these stars correlates with their variablilty. Since the period could be measured from Earth, astronomers could then take the star's brightness and figure out how far away it was (since stars that are further away are dimmer, just like a flashlight looks much dimmer from across the room than it does in your hand). Hubble used this technique to prove that the Andromeda Nebula was actually an independent galaxy, 900,000 light-years away!

FIGURE 3-3:
Andromeda
Galaxy

(refer to page 278 for more information)

© 2001, www.arttoday.com

An Expanding Universe

Hubble began to measure the distances to many galaxies and compare them with the galaxies radial velocity. By the early 1930s, he had accumulated enough measurements to conclusively prove that the further away a galaxy was, the faster it was moving. The only explanation for this was that the entire universe was expanding! Hubble also showed that this expansion was the same in all directions. This realization doesn't mean that all the galaxies are running away from us in particular—the expansion would look the same if we lived in a different galaxy. Instead, there was suddenly proof that not only was the universe expanding, but that all its matter had started in a single point.

To visualize an expanding universe, imagine blowing up a balloon. If you draw a bunch of spots on a deflated balloon, then blow it up, all the spots will move away from each other at an equal rate. Now imagine this situation in the universe.

A Competing Theory Disproved

Not everyone believed the Big Bang theory. In the 1940s and 1950s, the main competing theory was the Steady State theory, proposed by Fred Hoyle. Hoyle actually made up the name Big Bang for the opposing theory to mock it. Unfortunately for him, the name stuck!

Hoyle's theory was that instead of having an origin around 15 billion years ago in the Big Bang, the universe had been around forever and was in a steady state. He did admit that the universe was expanding, but instead of having galaxies moving away from each other, he proposed that space was constantly being created in between galaxies, giving them the appearance of moving away. He further proposed that in this new space, matter was also created to keep the average density of the universe constant. This aspect of the theory would allow new galaxies to be created in between old ones.

Elemental Abundances

Two pieces of observational evidence favored the Big Bang theory over the Steady State theory. The first was the observed abundances of different elements. From measurements of the Sun, Earth, and other stars, astronomers came up with a table showing the relative amounts of each element present. When the observed amounts of light elements were compared to theories of what would have been created during the first few minutes of the Big Bang, they agreed very closely.

For example, the model predicted that in those early moments, enough helium would have been formed to make up about 25 percent of the total mass of the universe. When observations were made of the compositions of stars and gas clouds, this amount agreed with what was predicted. The Steady State theory was unable to explain the element abundances.

Cosmic Microwave Background

The second piece of evidence essentially disproved the Steady State theory permanently. In 1965, cosmic microwave background radiation was discovered. Think back to the last bits of energy from the primordial fireball of the Big Bang being stretched out to longer and longer wavelengths as the newly formed universe began to expand. In the 1960s, scientists began to try to calculate what wavelength this radiation would currently have and how it could be observed.

Two astronomers, Robert Dicke and Jim Peebles, realized that after its rapid expansion period ended, the universe would have had a temperature of about 3 Kelvin (that's only 3 degrees above absolute zero, where all motion stops). At this temperature, a body will give off radiation that is detectable at a variety of different wavelengths, but with the most radiation at a wavelength of about 7.35 centimeters. Dicke and Peebles began trying to design an instrument that could detect this radiation.

QUESTIONS?

How old is the universe?
Omega and the Hubble Constant are both necessary for calculating the age of the universe. Currently, the several methods of measuring the Hubble Constant all have different errors. For this reason, the age of the universe is usually given as about 15 billion years, but it could be as old as 20 billion or as young as 10 billion.

At the same time, two different astronomers were trying to measure the radiation given off by the Milky Way galaxy (our own galaxy). They happened to be observing at wavelengths around 7 centimeters. But no matter how they refined their techniques, they kept coming up with an annoying noise interfering with their observations. This noise seemed to be coming from all directions, and it had almost no variation. Arno Penzias and Robert Wilson had to write up a paper describing their failed experiment.

A few months later, they heard that the group led by Dicke and Peebles was in fact searching for an omnidirectional radiation source at about that wavelength! They pooled their observations with the theory of Dicke and Peebles, and eventually Penzias and Wilson won the Nobel Prize for their discovery. Their contribution was the single most important discovery confirming the Big Bang theory for the origin of the universe.

The Age of the Universe

The Big Bang model for the origin of the universe can also let us determine the age of the universe. This area of research is still quite active, and no definite conclusions have yet been reached. Two very important parameters are the mass density of the universe (called *Omega* after the Greek letter that symbolizes it in equations), and the rate of expansion (called the *Hubble Constant*). Once Omega and the Hubble Constant are known, we can estimate the age of the universe.

Omega

There is a mass threshold that will make the universe expand forever, eventually contract into a point, or do neither. This quantity is what Omega measures. It's actually the ratio of the amount of matter in the universe to the amount of matter needed to keep the universe from either expanding or contracting. If Omega is greater than one, there is enough mass to eventually stop the universe's expansion; at some point the universe will reverse its course and start contracting. The end of this contraction, sometimes called the Big Crunch, would lead to a closed universe.

If Omega is less than one, then the universe is an open universe, and the expansion of the universe will continue forever. If Omega is exactly one, the universe is called a flat universe or a critical universe, and it is balanced exactly between infinite expansion and collapse. In this case, the universe will never quite finish its expansion, but it won't contract either. This value is called *critical density*.

The Hubble Constant

The Hubble Constant, the rate of expansion of the universe, is expressed in terms of velocity per distance, and is usually given in kilometers per second per megaparsec (a megaparsec is about 3 million light years, or 1 million parsecs). This term is important because if we know the rate at which the universe is expanding, and we know the size of the universe, then we can figure out how long the universe has been expanding—therefore how old it is! Actually, the Hubble Constant isn't really a constant; due to gravity, the expansion of the universe is slowing down over time, but scientists believe they can account for this factor in their equations.

FIGURE 3-4:
Galaxy
NGC 4214,
one of the
many places
where star
formation
continues

(refer to page
278 for more
information)

Courtesy of NASA and The Hubble Heritage Team (STScI/AURA)

CHAPTER 4

Constellations:
Patterns in the Sky

Ancient people looked at the sky and saw a variety of mythological gods, heroes, monsters, and other assorted creatures and objects. Once you know the basic constellations, learn to use certain bright stars as guideposts to locate deep-sky objects. This chapter will help you learn the major constellations, including their mythology, their relationships to each other, and how to identify them in the night sky.

The Development of Mythology

Greek mythology was the cornerstone of modern constellation terminology. Around the seventh and sixth centuries B.C., star groupings were named for the animals they resembled; the mythology they came to represent had not yet been formulated. The myths and legends currently associated with the constellations were fairly well-established by the end of the fifth century B.C., and astronomy and mythology were intricately woven together thereafter.

One of the first categorizations of the stars, replete with their mythology, was the *Catasterismi* of Eratosthenes (276–196 B.C.). This work is the oldest known catalog of Greek constellations. Hipparchus also produced a star catalog in the second century B.C., and in it he ranked stars by their magnitude, meaning brightness. He described stars in terms of six different degrees of brightness, the brightest being first-magnitude stars. A refined version of this system is still used today.

QUESTIONS?

How did constellations come about in the first place?
Ancient cultures gazed up at the stars in awe, and saw reflections of their own culture's mythology, gods, and heroes in the patterns in the sky. These patterns also allowed the ancient peoples to remember certain stars and constellations better, helping them navigate at night when the Sun wasn't available

The next generation star catalog was written by a Roman astronomer, Ptolemy of Alexandria. Writing his *Almagest* in the second century A.D., Ptolemy grouped 1,022 stars into forty-eight constellations, the most comprehensive listing at the time. His listing was, among other things, one of Copernicus's primary sources 1,500 years later.

Stars in Your Eyes

There are currently eighty-eight defined constellations. Constellations are usually located by two coordinates: right ascension and declination. Right ascension, similar to longitude, is an angle measured in hours and minutes as an arc sweeps around Earth. Declination is much like latitude, measuring the distance of an object from the celestial equator. The following constellations are noted with these parameters, and one can use a star map for more help. Chapter 5 will help you understand these coordinates better and use them to locate stars and constellations in the sky.

Be aware that some constellations are visible only from the Northern Hemisphere, and some only from the Southern. The constellations you can see, then, depend on the latitude from which you are observing. Because Earth's inclination (tilt) changes with the seasons, some constellations visible in the summer in the Northern Hemisphere will be visible in the winter in the Southern Hemisphere, and vice versa. The best times to observe each constellation are also noted. (Refer to Appendix A for additional information.)

The constellations in this list are the major ones, and also some of the easiest to see:

ANDROMEDA, *the Princess*: An Ethiopian princess, and the daughter of Cepheus and Cassiopeia. Legend has it that Cassiopeia claimed to surpass the goddess Juno in beauty. As punishment for her pride and vanity, Cassiopeia had to sacrifice her daughter Andromeda to a sea monster. Andromeda was chained to a rock by the ocean, and left to die. Fortunately for her, Perseus happened along, fell in love with her on the spot, and married her (after killing Cetus the sea monster, of course).

Andromeda is a fall constellation in the Northern Hemisphere, and is best observed in October. She can also be seen in the spring in the Southern Hemisphere. Andromeda has a right ascension of 34 minutes, and declination of 39 degrees, 15 minutes.

AQUARIUS, *the Water Carrier*: Ganymede, a Greek cupbearer to the gods. Jupiter saw Ganymede and apparently decided that the beautiful youth was good cupbearing material. Jupiter turned himself into a bird and carried the young man back to Mount Olympus.

Aquarius is observed most easily in the fall from the Northern Hemisphere, and in the spring from the Southern Hemisphere. The constellation has a right ascension of 22 hours, 42 minutes and a declination of −10 degrees, 28 minutes.

ARIES, *the Ram*: Considered the first constellation of the Zodiac, it represents the Golden Fleece. The magical ram was used to carry Nephele's children, Phryxus and Helle, away from her husband Athamas's second wife, Ino. Phryxus survived the trip and sacrificed the ram to Jupiter; the ram then became the constellation Aries. The fleece of the ram was the Golden Fleece, the subject of a quest by Jason and the Argonauts.

Aries can be seen best in October in the Northern Hemisphere, and spring in the Southern Hemisphere. Aries has a right ascension of 2 hours, 41 minutes, and a declination of 22 degrees, 34 minutes.

CANCER, *the Crab*: Believed to represent the crab sent to foil Hercules during the second of his twelve labors. As Hercules was fighting the Hydra, the crab nipped at his heels and posed a general annoyance. Although Hercules ended up killing the crab, Juno decided it had tried hard enough, and immortalized it as a constellation.

Cancer is most visible in early spring in the Northern Hemisphere, and fall in the Southern Hemisphere. It has a right ascension of 8 hours, 30 minutes and a declination of 23 degrees, 34 minutes.

CAPRICORNUS, *the Sea Goat*: A constellation tied to several others. The story goes that the Greek god Pan was having a feast on the river Nile when he saw the giant Typhon approaching. In an attempt to hide, Pan jumped into the river and disguised himself; the part of his body that was underwater turned into a fish, and his upper body turned into a goat. When he later saw the giant Typhon fighting with Zeus, Pan scared the giant away, and was rewarded for his bravery with a constellation in the heavens.

Capricornus is seen clearly in the late summer, especially August, in the Northern Hemisphere, and winter in the Southern Hemisphere. This constellation has a right ascension of 12 hours, 3 minutes and a declination of −19 degrees, 21 minutes.

CASSIOPEIA, *the Ethiopian Queen*: Mother of Andromeda, wife of Cepheus. After Cassiopeia professed her beauty to be greater than that of any of the goddesses, Neptune (god of the sea) sent Cetus, a sea monster, to destroy her. Although Neptune did add Cassiopeia to the constellations in the sky, some say he had the last laugh because he placed her head pointing at the North Star, meaning she was seen as upside down for half the evening.

Cassiopeia is visible only from the Northern Hemisphere and is best observed in June. Since she's located near the North Star, however, she is visible all year. She has a right ascension of 52 minutes and a declination of 60 degrees, 18 minutes.

ALERT

Visible constellations change during different times of the year. When observing in December and January from the Northern Hemisphere, look for Orion—the great hunter is the brightest constellation in the winter sky. Orion is easily detected by the three stars in his belt.

CEPHEUS, *the Ethiopian King*: Father of Andromeda, husband of Cassiopeia. Cepheus does double duty as a constellation; he is related to Andromeda and Cassiopeia, and he was also a member of the band of Argonauts that traveled with Jason in search of the Golden Fleece.

Cepheus is also located near the North Star, and visible all year, but he is best observed in the summer from the Northern Hemisphere. Cepheus has a right ascension of 22 hours, 25 minutes and a declination of 72 degrees, 34 minutes.

CETUS, *the Sea Monster*: Perhaps the most famous sea creature of Greek and Roman times. He was sent to Ethiopia to deal with Andromeda, daughter of Cassiopeia and Cepheus, and ended up being killed by Perseus. His constellation is located a good distance from that of his thwarters, especially Perseus.

Seen best in October from the Northern Hemisphere and spring in the Southern Hemisphere, Cetus has a right ascension of 1 hour, 43 minutes and a declination of −6 degrees, 22 minutes.

CORONA BOREALIS, *the Northern Crown*: Usually connected with the crown of Ariadne, daughter of the Cretan King Minos. Minos and his queen had a son who turned out to be half bull and half man. Aptly called the Minotaur, the creature was locked in a maze designed especially for him. He dined on hapless Athenians, one of whom was supposed to be the hero Theseus. Ariadne took a liking to Theseus, and offered him a way out of the labyrinth in exchange for her hand in marriage. Theseus killed the Minotaur and escaped, but abandoned Ariadne. She ended up following the wine god Dionysus, and married him. In a burst of joy, she threw her crown into the sky, where its jewels became the constellation.

Corona Borealis is visible between latitudes 90 and –50 degrees, and is best observed in July from the Northern Hemisphere and in winter from the Southern Hemisphere. It is positioned with a right ascension of sixteen hours, and declination of 32 degrees.

QUESTIONS?

What is an asterism?
An asterism is a group of stars that's actually part of one or more constellations. The Summer Triangle, made up of Vega, Deneb, and Altair, is an asterism, as is the Big Dipper

CYGNUS, *the Swan*: Of varied origin. One of several possible myths suggests that Cygnus was the pet swan of Cassiopeia. Another places Cygnus as the son of the god Neptune, who turned him into a swan to save him from Achilles. A third myth identifies Cygnus as the musician Orpheus, who was murdered at Dionysus's bidding by a woman from Thrace. A fourth suggests that the constellation Cygnus represents the son of Sthenele. He supposedly became a constellation when his friend Phaëthon was unjustly killed by Zeus; he had a reconfiguring of sorts, and literally metamorphosed into a swan.

Cygnus is most easily seen in July from the Northern Hemisphere, or in winter from the Southern Hemisphere. This constellation has a right ascension of 20 hours, 36 minutes and a declination of 49 degrees, 35 minutes. The constellation Cygnus includes the bright star Deneb, which is part of the Summer Triangle asterism.

DRACO, *the Dragon*: Also has different creation myths. The most popular explanation is that Draco was the guard of the golden apples from Hercules' eleventh of twelve labors. The goddess Juno received them as a wedding gift, and left them with the daughters of Hesperus, to be guarded by Draco. Hercules, in retrieving these apples for Atlas, ended up killing Draco, and some say that Juno rewarded the dragon through this constellation. Another common story of the Draco constellation is that of Cadmus. He was sent to search Earth for his kidnapped sister Europa, who had been turned into the bull Taurus. Cadmus, unable to find her, followed the advice of Apollo and went ahead to build a new city, but all of his servants were killed by a ferocious dragon who Cadmus promptly killed.

Draco is a spring constellation and is most visible March through May. However, the constellation is circumpolar (meaning it travels in circles around the north celestial pole) and visible all year from the Northern Hemisphere and not at all from the Southern. He is positioned at a right ascension of 17 hours, 57 minutes and a declination of 66 degrees, 4 minutes.

ERIDANUS, *the River*: An actual river in Italy, today called the Po. Phaëthon, friend of Sthenele and son of Apollo and Clymene, wanted proof that he was genuinely descended from the mighty Apollo. He requested to drive Apollo's chariot of the Sun, and the god reluctantly agreed. Mere mortal that he was, Phaëthon was not strong enough to control the horses, and he ended up colliding with both Earth and the heavens. To avert a colossal disaster, Apollo sent a thunderbolt and essentially hurtled Phaëthon into the heavens. His body fell into the river Eridanus. This constellation is the longest in the sky.

Eridanus is seen most clearly in early winter, especially November from the Northern Hemisphere, or summer in the Southern Hemisphere. It has a right ascension of 3 hours, 55 minutes and a declination of –15 degrees, 50 minutes.

GEMINI, *the Twins*: Represents Castor and Pollux. Born to Leda (and brothers of Helen), the twins had different fathers; Jupiter came to Leda as a swan, and she was also with her husband (King Tyndarus) that evening. Pollux ended up immortal, Castor mortal; this myth was in keeping with the classical idea of all earthly twins consisting of one immortal. The brothers were both Argonauts in Jason's quest for the Golden Fleece, and were warriors of the Trojan War. The mortal Castor was killed by Lynceus, and Pollux was in such a state of despair that he asked Zeus to place the two of them together forever.

Gemini is located between Taurus and Cancer. The constellation is most visible in January from the Northern Hemisphere, and is a summer constellation in the Southern Hemisphere. It is positioned at a right ascension of 7 hours, 12 minutes and a declination of 22 degrees, 45 minutes.

HERCULES, *the Hero*: One of the all-time greatest heroes in Greek and Roman mythology. Perhaps he deserved more than one constellation! The son of Jupiter and Alcmene, and the immortal twin brother of Iphicles, he was constantly thwarted in life by his stepmother Juno. Hercules married Megara but killed both her and his children after being driven crazy by his stepmother. Guilt wracked Hercules, and he atoned for his sins by performing the Twelve Labors listed below. In the constellation, Hercules is shown wearing the Nemean Lion's skin, with Cerberus by his side and his foot on Draco's head.

Hercules' Twelve Labors

1. Kill the invulnerable Nemean Lion (Leo)
2. Slay the Lernaean Hydra
3. Capture the Cerynitian Hind, the golden-horned stag
4. Capture the Erymanthian Boar; slay two centaurs, Pholus and Chiron
5. Clean the Augean Stables by forcing two rivers to pass through them
6. Kill the Stymphalian, flesh-eating birds
7. Capture the Cretan Bull
8. Capture the flesh-eating mares of Diomedes
9. Steal Hippolyta's girdle
10. Steal the cattle of Geryon
11. Steal Cerberus, the three-headed dog
12. Acquire the Golden Apples for Atlas

Hercules is located between Bootes and Lyra, and is best observed in June from the Northern Hemisphere, or winter from the Southern Hemisphere. This constellation is positioned at a right ascension of 16 hours, 42 minutes and has a declination of 36 degrees, 28 minutes.

HYDRA, *the Water Snake*: Symbolic of Hercules' second labor, the Lernaean Hydra. This nine-headed monster was able to regenerate dismembered heads and was therefore quite difficult to kill. Hercules ended up having his cousin Iolaus burn the neck of each head he cut off, thereby enabling him to finish off the beast.

Hydra is most easily observed in the Southern Hemisphere in the fall, but can also be seen in the Northern Hemisphere between February and May. It has a right ascension of 10 hours, 12 minutes and a declination of −19 degrees, 33 minutes.

FACTS

The brightest star in the night sky is Sirius. This star is twenty times brighter than the Sun (but the Sun looks much brighter because it is much closer to us). Sirius is sometimes called the dog star because it is located in the Canis Majoris (big dog) constellation.

LEO, *the Lion*: The unfortunate first of Hercules' twelve labors. He was supposed to have a skin that could not be cut or broken, so Hercules avoided this issue by strangling the lion to death. He then proceeded to remove the lion's skin using one of its own claws.

Leo is best observed in March from the Northern Hemisphere, or in the fall from the Southern Hemisphere. This constellation has a right ascension of 20 hours, 36 minutes and a declination of 49 degrees, 35 minutes.

LIBRA, *the Scales*: Generally thought to represent balance and justice. Libra is usually associated with Astrea, the goddess of justice, who weighed the good and evil of humans on her golden scales. It is sometimes said that Astrea became frustrated with the evil of man and returned to the heavens, leaving in such a hurry that the golden scales remained on Earth. The Romans supposedly created this constellation so that Astrea could have her scales back. Virgo, the winged virgin, is often represented as Astrea.

Libra is located next to the constellation Virgo, near Virgo's hand. This constellation is most prominent in May in the Northern Hemisphere, and in the winter in the Southern Hemisphere. Libra is positioned at a right ascension of 15 hours, 12 minutes and a declination of –15 degrees, 43 minutes.

LYRA, *the Lyre*: The instrument of Orpheus, son of Apollo and Calliope. Orpheus was extremely skilled with his lyre, producing music that had a soporific effect on beasts and plants alike. He married Eurydice, but she was killed by a snake shortly thereafter. Orpheus, always the charmer, convinced Hades and Persephone to let him return Eurydice to the realm of the living, but she was not strong enough for the journey. The women of Thrace sought Orpheus as a mate; he apparently failed to return their interest and was dismembered. Zeus then hurled his lyre into the sky, creating the constellation Lyra.

Lyra contains Vega (the second-brightest star in the sky), which is part of the summer triangle along with Deneb and Altair. Lyra is best seen in July from the Northern Hemisphere, and is visible in winter from the Southern Hemisphere. This constellation has a right ascension of 18 hours, 52 minutes and a declination of 36 degrees, 51 minutes.

ORION, *the Hunter*: Another constellation with multiple interpretations. One myth celebrates Orion as the son of Poseidon and Euryale, an Amazon queen and hunter. Receiving his mother's skill as a hunter, Orion proclaimed himself to be the world's greatest hunter, but was killed by a single scorpion sting. Another myth portrays Orion as a motherless gift to the gods who became a famous blacksmith and hunter. He longed to marry Merope, but the King of Chios refused to allow it; he blinded Orion, whose sight was later returned by Apollo. Orion then set his sights on Artemis, the goddess of the hunt, but her brother Apollo tricked Artemis into killing Orion. Orion's faithful hunting dog follows at his heels in the constellation Canis Major, which includes the bright star Sirius (sometimes called the Dog Star).

The constellation Orion is a large and detailed one, and includes many famous objects. The bright star Betelgeuse makes up Orion's left shoulder. Three other bright stars make up Orion's belt, a configuration that was recognized by many cultures (and which might be reflected in the layout of the three great pyramids in Egypt). Three fainter objects hanging down from Orion's belt make up his sword. The central object in the sword is in fact not a star, but the Orion Nebula, a famous and spectacular deep-sky object. The Horsehead Nebula is also nearby.

ESSENTIALS

Even though stars in a particular constellation appear to be close together, they may not be close at all. In fact, some could be much further from Earth than others. The constellations would look very different from a planet orbiting a star elsewhere in the galaxy.

Orion is best observed during the winter months, mainly December, in the Northern Hemisphere and during the summer in the Southern Hemisphere. He is readily identified by the three stars on his belt, and is positioned at a right ascension of 5 hours, 36 minutes, and a declination of 4 degrees, 35 minutes.

PERSEUS: A favorite Greek hero. Son of Zeus and Danae, he slew the Medusa, his most impressive feat. As the story goes, the Gorgon Medusa was difficult to slay because one glance at her would turn anyone to stone. Perseus enlisted the aid of several gods; Athena provided him with a shield, Hermes gave Perseus his winged sandals, and Hades gave him the helmet of invisibility. After killing the Medusa, Perseus held up her head to Atlas, turning him to stone, and Atlas increased in size until he became a mountain and could support the world. Perseus went on to rescue Andromeda, and became a constellation as reward for his many great deeds.

In the sky, Perseus is seen with the Medusa head in one hand and his sword in the other. This constellation is seen best in November from the Northern Hemisphere, and in the spring from the Southern Hemisphere. He is positioned at a right ascension of 3 hours, 40 minutes and a declination of 41 degrees, 45 minutes.

PISCES, *the Fish*: The forms that Venus and Cupid took to escape the mighty giant Typhus, the same giant that forced Faunus to metamorphose into a goat. Typhus was bent on destroying all the Olympian gods. Venus and Cupid disguised themselves as fish, and Minerva commemorated them with the Pisces constellation.

Pisces is an autumn constellation, seen best in September from the Northern Hemisphere and in the spring from the Southern Hemisphere. Pisces has a right ascension of 0 hours, 50 minutes and a declination of 11 degrees, 8 minutes.

SAGITTARIUS, *the Archer*: Generally represents Chiron, a centaur (half human, half horse) and Zeus's half brother. Chiron was an exceptionally gentle centaur, known for his skills in archery, education, and music. He supposedly tutored some of the Greek heroes, such as Hercules and Jason. Unfortunately, Hercules accidentally shot Chiron with a poisoned arrow. The noble centaur then offered to take the place of Prometheus in the underworld, and for his kindness he was immortalized in a constellation.

Sagittarius is seen in the Northern Hemisphere during the summer, especially July, and in the Southern Hemisphere during the winter. This constellation is positioned at a right ascension of 19 hours, 7 minutes and a declination of –25 degrees, 45 minutes.

SCORPIUS, *the Scorpion*: The creature responsible for Orion's death, possibly sent by Apollo, when Orion's boasting of his hunting skills grew too great. Scorpius is generally seen as dark, representing death and destruction.

Scorpius is placed in the sky as far from Orion as possible. Scorpius is visible in the summer months from the Northern Hemisphere, and in the winter in the Southern Hemisphere. This constellation has a right ascension of 16 hours, 17 minutes and a declination of –22 degrees, 59 minutes.

TAURUS, *the Bull*: Symbolic of Jupiter, who turned into a bull when he decided he wanted to marry Europa. The princess was under constant guard by her father's servants, so Jupiter decided to change himself into a member of her father's cattle. After seducing her, he transformed back into a god, but realized he could not marry her. Instead, he arranged her marriage to the King of Crete. The Hyades are a V-shaped grouping of stars that form the face of Taurus; these represent the women chosen by Jupiter to raise his son Bacchus. Jupiter placed them in the heavens as a reward for their service.

Taurus is observed in the Northern Hemisphere most clearly in the early winter months, and can be seen in the Southern Hemisphere in the summer. Taurus has a right ascension of 4 hours, 15 minutes and a declination of 18 degrees, 52 minutes.

URSA MAJOR, *the Great Bear*: Created from another mythological story involving Jupiter. The hunter Callisto appealed to Jupiter, who took the form of Diana (the goddess of the hunt) in order to talk to her, but ended up creating a son, Arcas. Jupiter's wife Juno, displeased with her husband's philandering, transformed Callisto into a bear. When Arcas nearly killed his mother by mistake, Jupiter turned him into a bear as well.

FACTS

You can use the constellations to find north without a compass. Follow the two stars at the edge of the bowl of the Big Dipper up to a bright star. That is Polaris, the North Star, and also the end of the handle of the Little Dipper. If you stand and face Polaris, you are looking north.

Ursa Major, part of which is sometimes called The Big Dipper, represents Callisto. The Big Dipper is one of the most recognizable star patterns in the Northern Hemisphere, and was recorded by many cultures. In other cultures, it was thought to resemble a plow, a wagon or cart, the thigh of a bull, and even the government in Chinese mythology! The middle star of the handle of the Big Dipper is the double star Mizar and Alcor, which is easily seen through binoculars and was used as a naked-eye vision test in some early cultures!

Ursa Major is visible in the Northern Hemisphere for most of the year as one of the north polar constellations, and is located at a right ascension of 10 hours, 43 minutes and a declination of 55 degrees, 24 minutes.

URSA MINOR, *the Little Bear*: Symbolic of Arcas, who nearly killed his mother when she had been transformed into a bear by Juno. Jupiter transformed Arcas into a bear as well, and this Little Dipper grouping of stars was Arcas's reward from Jupiter (the dipper that we're all familiar with is actually the tail and part of the bear's midsection). Juno was supposedly so enraged at the celebration of Callisto and Arcas that she persuaded Neptune, god of the sea, to never allow the two bears to bathe in the ocean. As a result, both are circumpolar constellations, meaning they never appear to sink beneath the horizon. Circumpolar stars travel in circles around the north celestial pole so that they are visible all year round from the Northern Hemisphere. Cassiopeia, Cepheus, and Draco are also circumpolar constellations.

Ursa Minor is visible in the Northern Hemisphere all year. He is positioned at a right ascension of 14 hours, 46 minutes and a declination of 74 degrees, 20 minutes. Polaris, the pole star of the Northern Hemisphere, is part of this constellation; it can be located by tracing a line straight from the pointer stars of Ursa Major.

VIRGO, *the Virgin*: Sometimes synonymous with Astrea, the goddess of justice. She is usually seen with two ears of wheat, the brighter of which is the star Spica (Latin for wheat). Her scales appear in the Libra constellation. According to the legend, Jupiter sent Pandora down to Earth. She opened the one box she was told not to, and all the evil in life flooded out. Although Hope did not escape, the gods slowly left Earth to return to the heavens, and Virgo was the last to go.

Virgo is visible in the spring in the Northern Hemisphere, best seen in May and June, and can be seen in the Southern Hemisphere in the fall. She has a right ascension of 13 hours, 12 minutes and a declination of –3 degrees, 45 minutes.

Chapter 5

Navigate the Night Sky

Now that you know some of the constellations in our galaxy, how do you find them? First you need a coordinate system. Regular maps use the coordinate system of latitude and longitude. This chapter covers the various celestial coordinate systems, and defines terms such as meridian and zenith, altitude and azimuth, and right ascension and declination.

Terrestrial Latitude and Longitude

Before you try to read the sky, you need to understand how your position on Earth affects what you can observe above you. Earth is always moving, so latitude, longitude, seasons, and time of day (or night) collectively determine which part of the sky you'll see at a given time.

Latitude

Going back to Earth, what do latitude and longitude mean? Earth isn't really flat like a map, of course—it's a sphere, and we live on the surface. Latitude is actually the angle between a point on the equator, a point at the center of Earth, and a point at your location (let's consider Boston, in this example). Why does the equator come into this idea? For a coordinate system to make sense, it needs an origin. Coordinates start at the zero point. On Earth, we've defined the equator as the zero point for latitude. If you live on the equator, your latitude is 0 degrees. If you live north of the equator, say in North America, then your latitude is north, or positive; therefore Boston's latitude is 42 degrees north. If you were to draw a line from Boston straight in to the center of Earth, and then draw a line from the center to the equator, the resulting angle would be 42 degrees.

FIGURE 5-1:
Latitude ranges from 0 to 90 degrees

(refer to page 278 for more information)

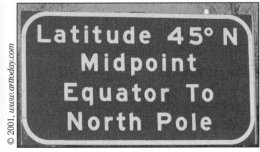
Latitude 45° N
Midpoint
Equator To
North Pole

© 2001, www.arttoday.com

If you live in the Southern Hemisphere, below the equator (say in South America), then your latitude is negative. Such latitude is sometimes referred to as south latitude. Just as 42 degrees north and +42 degrees both refer to Boston's latitude, if you live in Santiago, Chile, your latitude would be either 33 degrees south latitude, or –33 degrees. The maximum possible latitude would be either 90 degrees north or south. Ninety degrees north is the North Pole; if you draw an angle from the equator to the center of Earth to the North Pole, you get a right angle that has 90 degrees in it. Ninety degrees south is the South Pole.

Longitude

What about longitude? That concept is a little more complicated. The equator is a very obvious origin (zero line) to use when we're talking about latitude, but not for longitude! On Earth, the zero line for longitude is at the British Royal Greenwich Observatory, in Greenwich, England. This line is called the *prime meridian*, and was established there at an international conference in 1884.

ESSENTIALS

Start learning coordinates by figuring out where you are!

	LATITUDE	LONGITUDE
San Francisco, CA	37° 37′ N	122° 23′ W
New York City, NY	40° 39′ N	73° 47′ W
Sydney, Australia	33° 52′ S	151° 12′ E
Paris, France	48° 49′ N	2° 29′ E

Degrees of longitude proceed east and west of Greenwich, England, until they meet up at 180 degrees east and 180 degrees west out in the Pacific Ocean. The line where they meet is called the International Date Line, where it's a different day on either side of the line. To determine your longitude, take an angle between Greenwich, England, the center of Earth, and your location. For example, Washington, D.C., is at a longitude of 77 degrees west. The angle between the prime meridian, the center of Earth, and Washington, D.C., is 77 degrees.

FIGURE 5-2:
Globes display latitude and longitude

(refer to page 278 for more information)

© 2001, www.arttoday.com

Since a whole degree of latitude or longitude is quite large (a degree of latitude is about 69 miles, or 111 kilometers, long), degrees are subdivided into minutes and then into seconds. Each degree has 60 minutes, and each minute has 60 seconds (just like a clock). Then, each second can be divided even further into tenths, hundredths, or more. The exact coordinates of the U.S. Capitol in Washington,

D.C., are expressed as 38°53'23" N, 77°00'27" W. What these numbers mean is that the U.S. Capitol building is 38 degrees, 53 minutes, and 23 seconds north of the equator, and that it is 77 degrees, 0 minutes, and 27 seconds west of Greenwich, England.

The Celestial Sphere

Now that you know how to find latitude and longitude on Earth, you can use the same technique to talk about the positions of stars. First, you'll have to imagine a huge hollow sphere that surrounds Earth. Think of all the stars as dots painted on the sphere. Of course, we know this analogy isn't really correct—some stars are much further away than others—but the visual image presents a handy way of thinking about the sky.

FIGURE 5-3:
The celestial sphere

(refer to page 278 for more information)

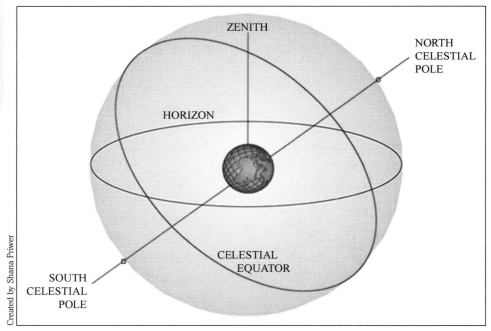

Created by Shana Priwer

The imaginary hollow sphere is called the *celestial sphere*. You can imagine that the celestial sphere rotates around Earth once every twenty-four hours. We know that it is actually Earth that rotates on its axis once every twenty-four hours, making the stars (as well as the Sun and Moon) appear to rise and set, but it's easier if we just imagine the celestial sphere as the object undergoing rotation. Since Earth actually rotates from west to east, we perceive the celestial sphere as rotating from east to west above us.

The Celestial Poles

Now that we've imagined the celestial sphere, we need some reference points in order to define a celestial coordinate system. First, the stars on the celestial sphere seem to rotate around the north and south celestial poles. These are the points on the celestial sphere that are directly above Earth's North Pole and South Pole. If you drew a line from the North Pole extending down through the center of Earth and then out the South Pole, then continued that line up and down in both directions until it intersected the celestial sphere, it would hit the celestial sphere at the north celestial pole and the south celestial pole.

FIGURE 5-4:
Star trails around Polaris

(refer to page 278 for more information)

© 2001, www.arttoday.com

Of course, the north and south celestial poles are just imaginary points in the sky, but luckily for those of us in the Northern Hemisphere, there's a fairly bright star that's very close to the north celestial pole! This star is Polaris, also sometimes called the North Star. If you imagined that you were looking down on Earth's North Pole, Earth would seem to rotate around it. Similarly, the stars on the celestial sphere seem to rotate around Polaris. If you take a long-duration exposure of the sky with a stationary camera pointed at Polaris, the stars around Polaris will all make circular streaks but Polaris will stay in the same place.

Another interesting fact about Polaris is that its height in degrees above the horizon is the same as your latitude! If you were in Boston, Polaris would be about 42 degrees above the horizon. If you were at the equator, Polaris would be right on the horizon, and if you were at the North Pole, Polaris would be at the top of the sky. Early explorers, especially those on boats who needed to keep track of their position, used this information. Interestingly, Polaris wasn't always the North Star. Earth's axis has wobbled and dipped with time, and this variation means that the north celestial pole has appeared to change places on the sky.

Celestial Equator

Another important imaginary line on the sky is the *celestial equator*. This line is actually an imaginary circle on the celestial sphere, directly above Earth's equator. It is always exactly 90 degrees from the celestial poles, just like Earth's equator is 90 degrees from its poles. As the celestial sphere appears to rotate above us, all the stars rotate on paths that are parallel to the celestial equator. At the North Pole, where Polaris and the north celestial pole are exactly overhead, the celestial equator is on the horizon.

Other Celestial Coordinates

Now that you have a better understanding of coordinates relative to Earth itself, you need to learn the terms that identify your own frame of reference. As you learned earlier, what you can observe is determined by your location on Earth, as well as time and seasons.

Horizon and Zenith

The *horizon* is the edge of our local sky. Since we are located on the surface of Earth, which is not transparent, we can see only half of the celestial sphere at one time. If you're in a place with lots of high mountains or tall buildings, you can see even less than half the sky. The point straight overhead on the celestial sphere is called the *zenith*. The zenith is the point you'll see if you lie on your back and look straight up at the stars. It is always 90 degrees from the horizon.

Meridian

If you draw an imaginary line on the celestial sphere (it's really an arc) that goes from the north point on the horizon up through the zenith and then down through the south point on the horizon, this line is called the *meridian*. Since the sky rotates from east to west above us, the meridian marks the halfway point. The meridian is also useful during the day, since it separates the morning and afternoon locations of the Sun. In the morning, the Sun is east of the meridian (because the Sun rises in the east). At local noon (assuming you're in the middle of your time zone and not observing daylight-saving time), the Sun is right on the meridian, and in the afternoon the Sun moves past the meridian, to the west.

QUESTIONS?

What do A.M. and P.M. mean?
In Latin, the morning is *ante meridiem*, which is abbreviated A.M. and means "before meridian." At noon, the Sun is on the meridian, and after noon, it is *post meridiem*, or P.M., which means "after meridian."

At night, as stars rotate through the sky on the celestial sphere, they start in the east, rise up to their highest point when they cross the meridian, and then descend and set in the west. Each star reaches its highest altitude, or height above the horizon, when it crosses the meridian. The angle that a star path makes with the horizon is 90 degrees minus the observer's latitude.

Depending on where the stars are, however, sometimes they never set. Polaris, for example, is always in the same place in the sky, and the stars rotate around it. Stars near Polaris are called circumpolar—over the course of the night, they travel a path that's a circle around the North Star, but never rise and set. The stars that are circumpolar at your location are those whose angular distance from the north celestial pole, or Polaris, is equal to or less than your latitude.

If you lived at the North or South Pole, all the stars would have this behavior—the whole sky would just turn in circles above you, but the stars wouldn't rise or set. This example also shows why some stars are never visible from some places in the Northern and Southern Hemispheres.

If you're in the Northern Hemisphere, then stars whose distance from the south celestial pole is less than your latitude will never be visible from your location.

The Ecliptic

The ecliptic is another imaginary line we can draw in the sky. Ancient astronomers tracked the motion of the Sun through the heavens, and determined that the Sun seems to drift in an easterly direction over the course of a year. It makes one full circuit through the sky in one year, or just about 1 degree per day. We know now that this apparent drift is caused by the motion of Earth in its orbit around the Sun.

Defining the Ecliptic

Let's go back to the celestial sphere model to see what the Sun's path does there. If we look at how the Sun seems to drift through the stars, the path that it travels is called the *ecliptic*. The ecliptic is a circular path that goes around the celestial sphere; this path is tilted at an angle of 23.5 degrees with respect to the celestial equator.

FACTS

The ecliptic path seems to be tilted at 23.5 degrees because it reflects Earth's rotation, and the axis is also tilted at a 23.5-degree angle.

The ecliptic crosses the celestial equator in two places, which are called the *spring* and *fall equinoxes*. The ecliptic is really the projection of Earth's orbit onto the celestial sphere, just like the celestial equator is the projection of Earth's equator onto the sphere. Remember that the Sun rises and sets each day because of the rotation of Earth, but it moves through the heavens over the course of a year because of the orbital motion of Earth around the Sun.

Relation to the Zodiac

Early astronomers knew that the ecliptic was special because it tracked the Sun's motion. The study of astrology, in fact, is based on just this motion. The twelve constellations that lie along the ecliptic on the celestial sphere are the signs of the zodiac! Your zodiac sign is the constellation that the Sun was in at the time of your birth. If you were born in May, for example, the apparent position of the Sun at the time of your birth was in the constellation Aries.

Bear in mind that horoscopes are based on a 4,000-year-old system, and due to the wobble of Earth's rotation the position of the constellations have changed over the last 4,000 years. So even if the Sun is in Aries on your birthday now, your sign might still be Taurus from the old system!

The ecliptic is also special because it's not only the plane of Earth's orbit, it's also the plane of the solar system. Why should we care about this phenomenon? Not only are the signs of the zodiac all located along it, but it's also the location of all the planets when they're visible in the sky. If you're looking for a planet, it will be located somewhere along the ecliptic (assuming the planet is currently visible, of course). If a few planets are visible at the same time, they'll all line up in the sky along the ecliptic! This scenario is called a *conjunction*.

Altitude and Azimuth

Now that you know how the sky moves, you're ready to start trying to find individual stars. The first way of specifying a star's position is what's called a *local position*. Local position is similar to giving someone directions to the store by saying, "go down the street a mile to the north, then turn to the west and go three blocks." It's a perfectly good description of the location of the store assuming that everyone who wants to go there starts at your house. Astronomically, this descriptive method is similar to the altitude-azimuth (alt/az) system:

The **altitude of a star** is defined as how many degrees it is above the horizon. A star on the horizon would have an altitude of 0 degrees, and a star at the zenith would have an altitude of 90 degrees because it's directly overhead.

The **azimuth of a star** is defined as how many degrees along the horizon it is from north. A star that's directly north would have an azimuth of 0 degrees, east is 90 degrees, south is 180 degrees, west is 270 degrees, and the coordinates increase up to 360 degrees, then turn back over to zero at true north again.

Altitude and azimuth are very easy to understand, but they're good only for people in your exact location at the exact time you're looking. Since stars move through the sky over the course of the night, their altitude and azimuth change with time as well.

Right Ascension and Declination

A more general coordinate system is called the *equatorial coordinate system*, used by both professional and amateur astronomers. This coordinate system is more like giving someone the latitude and longitude of the store down the street, and letting him or her figure out how to get there. Because these coordinates do not change with time and location, they are used in star maps, books, magazines, and computer programs.

Right Ascension

Remember the lines of longitude on Earth that run north-south from pole to pole? When we project those lines onto the celestial sphere, we get lines of right ascension. Before the age of digital watches and other accurate timepieces, stars were used to measure time, so right ascension (abbreviated RA) is given in hours, minutes, and seconds. RA increases as you go east across the sky. If you watch two stars whose coordinates are one hour of RA apart, the second star will cross the meridian an hour after the first star, and the second star will also rise and set an hour after the first one (unless either is circumpolar).

The zero point for longitude was arbitrarily chosen to be in England, and the zero point for RA was also arbitrarily chosen. The zero point of

RA is defined as the point on the celestial sphere where the Sun crosses the celestial equator at the spring equinox. Remember that this is the point in the spring sky where the ecliptic crosses the celestial equator. Just as lines of longitude on Earth converge at the poles, lines of RA converge at the celestial poles.

FACTS

There are 24 hours of RA and 360 degrees across the celestial sphere, so each hour of RA is equal to 15 degrees on the celestial sphere (or 15 degrees of Earth's rotation).

Declination

We can also take the lines of latitude on Earth, which are parallel to the equator, and project those onto the celestial sphere. There, they become lines of declination. Just as latitude on Earth is measured in degrees away from the equator, with positive for north and negative for south, declination (abbreviated dec) is measured in degrees away from the celestial equator, with positive degrees for stars or other objects that are north of the celestial equator and negative degrees for stars or other objects that are south of the celestial equator. Just as on Earth, objects that are exactly on the celestial equator have a dec of 0 degrees, objects at the north celestial pole are at dec +90 degrees, and objects at the south celestial pole are at dec –90 degrees.

The big advantage of the RA and dec system for locating stars is that these positions do not change over the course of a night. Unfortunately, however, the position of a star is dependent on the position of the north celestial pole and the celestial equator, and these do change slowly with time. The RA and dec of a star will change by about 1.4 degrees every 100 years. Because RA and dec change with time, star charts and catalogs must be drawn for a certain epoch, or time period. Most modern star charts are compiled with different positions every fifty years or so, with the most current being from the year 2000. If you're using an old star atlas, make sure to check the date, or your star positions could be off by a few degrees!

Finding Your Way

Once you know the altitude and azimuth of the object you're looking for, you can go outside to try to find it. First, stand in a nice dark spot with a good view of the horizon all around (not too many trees or buildings in the way). Next, find north—use a compass, or follow the pointer stars in the Big Dipper's bowl up to Polaris.

Once you've got your bearings, start with your azimuth. Let's say you're trying to find an object with an azimuth of 225 degrees and an altitude of 30 degrees. Remember that north is 0 degrees azimuth. Directly behind you, or south, is 180 degrees azimuth. If you hold out your left arm, that'll be pointing directly west, which is 270 degrees. The direction you want, then, is halfway between your left arm and directly behind you. Turn and face in that direction. You can also just find the 225-degree mark on your compass and turn in that direction.

SSENTIALS

Converting from RA/dec to alt/az is useful if you want to see if a particular celestial object is visible to you. If the altitude is negative, that means it's currently below the horizon, and isn't currently visible from your location.

To find an altitude of 30 degrees, remember that the horizon is 0 degrees, and the zenith is 90 degrees. So, 30 degrees is one-third of the way up the sky. You can estimate the altitude in a number of ways. An easy way to do this is to hold your fist out at arm's length in front of you; at this distance, the average fist is about 10 degrees. Look straight in that direction, and see if you can find your desired object!

If you want a more precise way of measuring altitude, you can remember that the full Moon and the Sun both cover about 0.5 degrees in the sky, and that the pointer stars in the bowl of the Big Dipper are about 5 degrees apart.

CHAPTER 6

The Inner Solar System

The four planets of the inner solar system (Mercury, Venus, Earth, and Mars) are sometimes called the *terrestrial planets*, or *Earthlike planets*. These bodies are somewhat similar to our home planet, but also surprisingly different. Mercury can be unbearably hot, Venus has clouds of sulfuric acid, and Mars is very cold with a thin atmosphere.

Mercury

FIGURE 6-1:
The planet
Mercury

(refer to page
278 for more
information)

Courtesy of NASA/Jet Propulsion Laboratory (JPL), Caltech

Mercury, the innermost planet, is a small, scarred object that lies very close to the Sun. Mercury is larger than Pluto, but smaller than all the other planets. It's even smaller than two moons of the outer solar system, Ganymede (a moon of Jupiter) and Titan (a moon of Saturn), but it is denser than either of those and has more mass. Mercury's high density suggests that it has a large iron or metal core, surrounded by a relatively small rocky layer. Mercury has no natural satellites or moons.

Mercury is locked in an orbital resonance with the Sun, meaning it rotates on its axis three times for every two times it orbits the Sun. The result is that three Mercury "days" equal two Mercury "years." Mercury's orbit is also highly eccentric, meaning that it is far from a perfect circle. At its closest approach, Mercury is about 46 million kilometers from the Sun, but at its farthest distance it is about 70 million kilometers away. Mercury is so close to the Sun that the sunlit side (the day side) gets extremely hot, reaching temperatures of about 700 degrees Kelvin. Mercury's atmosphere is very thin, however, and can't insulate the dark side of the surface very much at all. The temperature on the night side of Mercury drops very quickly to as low as 90 degrees Kelvin. This temperature range is the most extreme in the entire solar system.

Mercury has been visited by only one spacecraft, *Mariner 10*, which made three flybys in 1974 and 1975. This mission was able to map only less than half of Mercury's surface. The images show that Mercury is covered with old impact craters, and its general appearance is quite similar to that of Earth's Moon. The surface also has large ridges that cut through some of the craters and other surface features. Scientists have interpreted these ridges, which can be hundreds of kilometers long and up to 3 kilometers high, to be the result of compression from when the planet cooled and shrank.

There is also some evidence of volcanism on Mercury, because some of the smooth plains appear similar to the maria on (see page 75) Earth's Moon. The lunar maria are known to be due to volcanic flooding; however, the smooth plains on Mercury could be due to debris thrown up by impact craters.

The planet Mercury bears no relationship to the metal by the same name. This liquid metal (mercury) is also known as *quicksilver,* the stuff inside your thermometer that expands to indicate your body temperature.

Unfortunately, Mercury is too close to the Sun to be safely imaged by Hubble Space Telescope, so it has been very difficult to see if there are any signs of volcanic activity or other odd features on Mercury's unseen hemisphere. Some images of Mercury have been taken by radio telescopes on Earth using radar, and have revealed a few anomalous spots on the other hemisphere that could possibly be large volcanoes.

We won't know for sure whether Mercury had any volcanic activity, or whether there are other exotic features on the unknown hemisphere, until a spacecraft mission returns to Mercury. Fortunately a new mission to Mercury, called the MESSENGER (Mercury Surface, Space Environment, Geochemistry, and Ranging) mission, is in the planning stages at NASA. MESSENGER is currently scheduled for launch in 2004, and would orbit Mercury starting in 2009. This mission would take images of the entire surface of Mercury, as well as study its composition with a variety of instruments.

Venus

Venus, the second planet from the Sun, is the most Earthlike, and is sometimes called Earth's sister planet. Venus was named after the Roman goddess of love and beauty, and it's the brightest object in the sky after the Sun and Moon. Venus is also called either the Morning Star or the Evening Star because it is often the first object visible at night, or the last one visible before dawn.

At first glance, Venus seems to deserve its name. Its thick atmosphere means that the surface is shrouded in bluish-white clouds—the solid surface is not visible at all from orbit. Early astronomers were able to see phases on Venus (since it's closer to the Sun than Earth is), and such phases observed by Galileo helped support Copernicus's heliocentric view of the solar system. You can see the phases of Venus yourself using a small telescope. You may also notice that Venus has no natural satellites, or moons.

FIGURE 6-2:
Volcanic features on Venus from the *Magellan* spacecraft

(refer to page 278 for more information)

Courtesy of NASA/Jet Propulsion Laboratory (JPL), Caltech

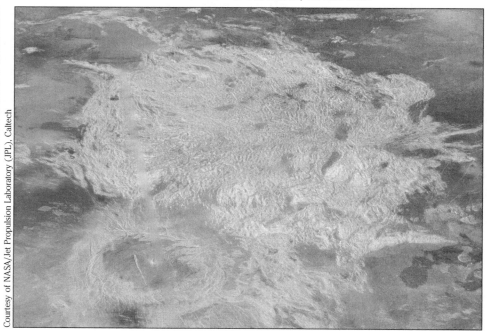

Even from above the clouds, Venus appears similar to Earth. Its diameter is about 95 percent of the diameter of Earth, and its chemical composition is also quite similar to Earth's. Before the Space Age, scientists and science fiction writers alike guessed that there might be life below the clouds of Venus, perhaps even jungles and whole civilizations. Unfortunately, spacecraft exploration of Venus has revealed a planet that is significantly less hospitable.

Why We Can't Live There

The atmosphere of Venus is very thick, and the pressure at the surface of Venus is ninety-two times the surface pressure at sea level on Earth (or ninety-two atmospheres). The atmosphere of Venus is mostly made up of carbon dioxide, but there are also clouds of sulfuric acid droplets. Light from the Sun comes in and warms up the surface, but the thick atmosphere keeps the leftover heat from being radiated out again. This heat entrapment causes the surface to get very hot. Scientists call this phenomenon a *runaway greenhouse effect*—it's an example of what happens in a glass greenhouse, in which the glass lets in the light from the Sun, but keeps the resultant heat from radiating out.

FACTS

The greenhouse effect on Venus results in a very high surface temperature, over 740 degrees Kelvin (900 degrees Fahrenheit), which is hot enough to melt lead! The surface of Venus is actually hotter than the surface of Mercury, even though Mercury is much closer to the Sun.

On Venus, the entire surface is covered with clouds, in contrast to Earth where we can see at least some of the surface at any given time. Fortunately, scientists discovered that long-wavelength radar waves could travel through the thick atmosphere of Venus, bounce off the surface, and return to a spacecraft or telescope. Called *radar mapping*, this technology allows us to build up images of the surface using radar instead of visible light. The *Magellan* spacecraft was specially built to produce radar maps of the surface of Venus.

The Landscape of Venus

Since the surface of Venus is so hot, there isn't any liquid water to help form the landscape. The surface of Venus is covered with mountains and volcanic plains. The volcanoes and lava flows cover about 85 percent of the surface of Venus. There are huge lava flows, up to hundreds of kilometers long, that have flooded the flat areas. There are more than

100,000 small volcanoes on Venus, and hundreds of larger ones. There are also odd circular volcanic features such as coronae.

In addition to mountains and volcanoes, Venus also has impact craters. The thick atmosphere of Venus causes small objects to burn up and be destroyed before they hit the surface, so there aren't any small craters (less than 2 kilometers in diameter) on Venus as there are on the Moon or other airless bodies. Some of the craters on Venus are surrounded by odd halos that could be caused by the winds blowing around ejected material. There are even lava flows that look as if they're coming out of some craters—the surface temperature on Venus is so high that the rocks are pretty close to melting already, and the added heat from an impact could melt them completely!

QUESTIONS?

Why is radar mapping the best option for learning about Venus's landscape?
We can't see through the dense atmosphere of Venus, and high temperature and atmospheric pressure make it hard for spacecraft to survive on the surface of Venus. Radar mapping gives us the best chance to "see" Venus's surface without fighting its hostile atmosphere.

The surface of Venus also has ridges and wrinkles that could be due to some kind of tectonic activity, perhaps similar to the events that cause earthquakes and plate tectonics on Earth. One of the big controversies about Venus is that the whole surface seems to be close to the same age, around 500 million years old. This scenario is very different from places like the Moon or Mars, where there are old, cratered areas and then much younger, more recently active areas. Some scientists think this dating information could mean that the whole planet of Venus went through a global resurfacing event 500 million years ago, and hasn't been nearly as active since then.

Earth

FIGURE 6-3:
The planet
Earth, as
seen from
the *Apollo 17*
spacecraft

(refer to page
278 for more
information)

Courtesy of NASA/Johnson Space Center [JSC]

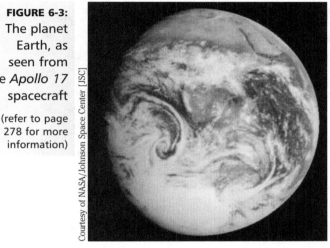

Earth, the planet we call home, is the most hospitable planet for life in the solar system—at least for terrestrial life. It's possible that life surviving under very different conditions thrives somewhere else in the solar system, but we have yet to find evidence of such life.

Earth is the only planet in the solar system where liquid water is stable at the surface. It's also one of the few solar system bodies that is currently geologically active (Io and Triton are two large satellites of the outer solar system that are also currently active, and other bodies may have current or recent geologic activity as well). Earth's crust is constantly recycled through volcanic eruptions and plate tectonics, which creates and destroys crust on the ocean floors. Water accounts for much of the geologic change on dry land as well. Erosion has carved huge channels such as the Grand Canyon, and snow in the form of glaciers has smoothed the areas near the poles. Wind erosion is also important in drier areas.

From orbit, Earth's difference from the other planets is obvious. Most of Earth is covered with water—either liquid water in the form of the oceans, solid water in the form of glaciers and polar ice caps, or vaporized water in the form of clouds. Peeking out from all this water are large landmasses that are green and brown, depending on the season. Earth's surface is geologically very young, and is constantly changing—due to humans on a small scale, and geologic forces on a much larger scale.

Our Moon

Earth is also unlike most other planets in that it has a very large moon. The Moon is a sizable fraction of the diameter of Earth, and it is almost

FIGURE 6-4:
A close-up
view of the
Moon's
surface

(refer to page
278 for more
information)

Courtesy of NASA/Johnson Space Center [JSC]

completely tidally locked to Earth. Tidal locking means that one side of the Moon always faces Earth. The Moon takes 29.5 days to orbit Earth, and that's also how long it takes the Moon to turn once on its axis. The far side of the Moon is sometimes called the *dark side* even though it's not actually dark; in fact, during what's called a *new Moon*, this side is brightly lit with sunlight! The Moon's far side always faces away from Earth, and was first seen by the Soviet *Luna 3* spacecraft in 1959. Because the Moon's orbit isn't perfectly circular, we can see a little bit more than one hemisphere of the Moon from Earth— the noncircular orbit causes a slight wobble (called a *libration*) of the Moon, letting us see a few extra degrees on each side.

Earth-Based Views and Effects of the Moon

The gravity of the Moon raises tides on Earth. The Moon tugs on the side of Earth facing it more than it tugs on the side facing away from the Moon, making tides fluctuate throughout the day. The Moon tugs on the solid parts of Earth as much as it tugs on the fluid oceans, but affects water more because water's much more easily deformable. The moon's gravitational influence explains why we have high tides and low tides about twice a day at the ocean coasts. The actual height of the tides at the coasts, however, is also dependent on underwater topography and weather, and is much more complicated than this simple explanation.

FACTS

Scientists think that the Moon might have been formed when a Mars-sized object crashed into Earth. This collision threw debris into orbit around Earth, and some of that debris might have coalesced into the Moon.

From Earth we can see phases of the Moon, which occur when different parts of the Moon are lit by the Sun during the Moon's approximately month-long orbit around Earth. A full Moon is visible when the entire Earth-facing side of the Moon is lit by sunlight, and a new Moon means only the side of the Moon facing away from Earth is lit by the Sun.

QUESTIONS?

What does *tidally locked* mean in reference to the Moon?
The Moon and Earth both have gravitational forces that attract each other. This gravitational force, in conjunction with the time it takes the Moon to revolve on its axis, means that, basically, the same side of the Moon is always facing Earth.

Through a small telescope, one can see that the Moon is covered with craters. Scientists first thought these craters were volcanic, because volcanoes cause most craters on Earth. Further study, however, showed many of the Moon's craters to be impact craters caused by space objects such as comets, asteroids, and meteors hitting the surface of the Moon. Unlike Earth and Venus, the Moon doesn't have enough atmosphere to stop smaller objects from making it to the surface. Craters on the Moon range from very large, thousands of kilometers in diameter, to seemingly microscopic, made by tiny grains of dust.

Surface Geology

Although craters are found almost everywhere on the Moon, some parts have more craters than others. From Earth, you can see dark and bright patches on the Moon's surface. These patches make up a configuration that vaguely resembles a face (the Man in the Moon), a rabbit, or whatever else your imagination can create. The dark patches are called *maria*, the Latin word for oceans. We know now that there is basically no water on the Moon (although there may be some ice in deep craters at the Moon's poles), and that the dark maria which cover about 16 percent of the surface of the Moon are in fact ancient lava flows. The bright parts of the Moon are called the *lunar highlands*, and they are much older and more heavily cratered than the maria. Most of the maria are concentrated on the near side of the Moon, facing Earth. This asymmetric arrangement of the Moon's mass may have helped it become tidally locked to Earth.

Despite the evidence of volcanic activity in the past, the Moon is currently geologically dead. The maria are younger than the lunar highlands, but most of the rocks on the surface of the Moon are between 3 and 4.6 billion years old. For the last few billion years the surface has mostly been collecting impact craters, which have produced a layer of rubble and debris, called a *regolith*, at the surface. This regolith layer is anywhere from a few meters to tens of meters thick.

Visits to the Moon

Before the Moon landings, there were a variety of theories for the formation of the Moon. These theories included such ideas as co-accretion (where Earth and Moon formed together out of a cloud of gas and dust in the early days of the solar system), capture (where the Moon formed elsewhere in the solar system and was captured into orbit around Earth by Earth's gravity), and fission (where the Moon split off from Earth early in its history because of excess rotation). However, none of these theories could adequately explain why the Moon has a much lower density than Earth due to its lack of metals, or why the Moon is depleted in volatile elements such as water.

 **SSENTIALS**

Lunar astronauts studied the Moon's surface, and brought back valuable samples of Moon rocks and dust. These rocks have been of great interest to scientists, and they are still being studied in laboratories around the world. Moon rocks have allowed us to learn more about the origin and early history of both the Moon and Earth.

Based on careful study of Moon rocks and other evidence, scientists came up with a new theory of the formation of the Moon, called the *giant impact theory*. In this theory, a large Mars-sized object impacted Earth early in its history, and threw out a disk of material that orbited around Earth. Most of this material fell back onto Earth, but some stayed in orbit long enough to bind together into larger chunks of rock, eventually forming the Moon. The heavier elements, like metals, fell back to Earth first, perhaps explaining why the Moon has much less metal than Earth. In addition, it would have been too hot for volatile elements like water to condense onto the forming Moon. This theory has not yet been proved, but it seems to be the best of the current ideas.

Mars

FIGURE 6-5:
The "Face on Mars" is just a strangely shaped mountain

(refer to page 278 for more information)

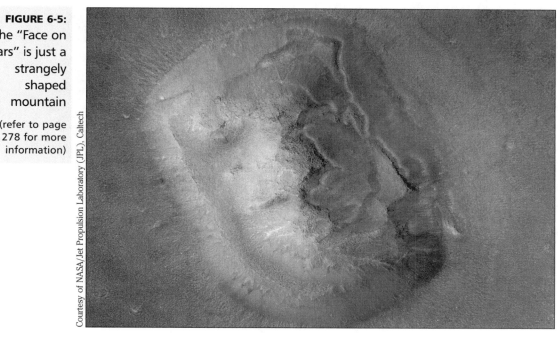

Courtesy of NASA/Jet Propulsion Laboratory (JPL), Caltech

Mars, the fourth planet from the Sun, is located between Earth and Jupiter, and is named after the Roman god of war. From Earth, Mars appears reddish; this apparent tint is why it's often called the Red Planet. Mars appears quite bright in the night sky, and when it is near opposition (its closest approach to the Sun), it is one of the brightest objects. Mars has been an object of speculation ever since early telescopes revealed seemingly linear features on its surface. Scientists and science fiction writers alike imagined them to be canals, perhaps the results of a vast irrigation system constructed by a race of Martians whose climate was becoming too cold and dry. Variations in the color patterns on the surface of Mars were interpreted as vegetation, perhaps crops or other plants that changed with the seasons.

Unfortunately, closer examination of Mars through spacecraft observation showed that the linear features seen by astronomers were probably just optical illusions. Interestingly enough, however, the most reliably seen canal corresponds to Valles Marineris, which is the largest canyon system on Mars. The changes in color patterns turned out to be windblown sand from the huge sandstorms that take place on Mars.

Similarities to Earth

Spacecraft observations have shown that Mars is a fascinating planet, and also quite Earthlike. If Venus is too hot for life, Mars is too cold—its average surface temperature is about −55 degrees Centigrade (−67 degrees Fahrenheit). Mars is farther from the Sun than Earth is, but its atmosphere is also very thin. The average surface pressure is only 7 millibars, which is less than 1 percent of the atmospheric pressure at sea level on Earth. The atmosphere is mostly made up of carbon dioxide, with a little nitrogen, argon, oxygen, and water.

Mars has polar ice caps similar to those on Earth, but the Martian ice caps are primarily carbon dioxide ice, or dry ice. In the summer, the ice cap in Mars's northern hemisphere heats up and the solid carbon dioxide turns into gas, leaving behind just a thin layer of water ice. This process raises the atmospheric pressure substantially, up to 25 percent, but the atmosphere is still much too thin to breathe (and has too little oxygen). Mars also has two small moons, Phobos and Deimos, which are thought to be captured asteroids.

Mars Exploration

Many spacecraft, both orbiters and landers, have observed Mars. The first observations of Mars came from the U.S. *Mariner 4* orbiter spacecraft in 1965. The first spacecraft to land on Mars was the Soviet *Mars 2* spacecraft, which made a crash landing in 1971. The two U.S. *Viking* spacecraft made soft landings on Mars in 1976. This very successful mission also included two orbiting spacecraft, and thoroughly characterized Mars. The landers returned 4,500 pictures and the orbiters returned over 50,000 pictures, mapping 97 percent of the surface of Mars.

The surface of Mars appears red due to the iron oxides that cover most of the surface. Even the sky is tinted pink (rather than blue) due to the different scattering properties of dust particles in the Martian atmosphere. The surface of Mars is covered with a variety of geological features. Some look quite familiar from Earth, such as large fields of sand dunes produced by the wind. The winds on Mars also cause huge sand storms, which can block the view of the surface for weeks or months. Mars contains some spectacular mountains, including the giant volcano Olympus Mons. This volcano is actually the largest mountain in the solar system. It is 24 kilometers high (78,000 feet), well over twice the height of Mt. Everest! Mars also has a huge canyon that puts Earth's Grand Canyon to shame: Valles Marineris is a system of canyons 4,000 kilometers long and 2 to 7 kilometers deep. Valles Marineris resulted from the stretching and cracking of Mars's crust during the formation of the Tharsis bulge, the volcanic rise where Olympus Mons is located.

ESSENTIALS

Although Earth is probably the only place in the inner solar system with liquid water, Mars may not be the only planet with frozen water. Ice could exist in craters at the poles of both Mercury and the Moon, according to recent data. The bottoms of these polar craters are in constant shadow from the crater rims.

Other canyons on Mars, however, appear to have been formed by some sort of fluid, in the way river valleys are carved by water on Earth. Water is the most obvious suggestion for what has eroded the surface of Mars, but there are alternate possibilities, including carbon dioxide flows and other, more exotic, suggestions. There is also evidence of giant outflow channels caused by immense floods. Some scientists even believe that the entire low northern hemisphere of Mars was once filled with a giant ocean of water. This flooding could have happened during a warmer, wetter period on Mars, when the atmosphere was thick enough to support liquid water on the surface. At that point in its proposed early history, Mars would have been much more Earthlike than it is now.

Over time, the carbon dioxide keeping the atmosphere thick and warming the surface through the greenhouse effect (as on Venus) would have been turned into carbonate rocks. This scenario occurred on Earth as well, but remember that Earth has plate tectonics, which recycle these rocks back into the crust and release the trapped carbon dioxide. Unfortunately, Mars has no plate tectonics, so it would not have been able to recycle its carbon dioxide. Eventually the atmosphere became thinner, and the surface colder.

Life on Mars

Scientists who believe in a warm, wet period in Mars's history think that much of this water could still be present there, possibly buried beneath its surface. This idea has important implications for the possibility of life on Mars, and also its future colonization by humans. If Mars once had much more water than it does now, it is quite possible that life could have evolved on Mars as it did on Earth.

FIGURE 6-6:
The surface
of Mars

(refer to page
278 for more
information)

Courtesy of NASA/JPL/Caltech

We're still not quite sure how life originated here on Earth, but life in general seems to require three things: liquid water, a sufficient source of energy, and certain chemical compounds including carbon and other elements. It is possible that early on, the Red Planet possessed all three of these requirements. If so, life could have originated on Mars completely separate from Earth, or it's possible that early life was transferred back and forth from Mars to Earth and from Earth to Mars through meteorites.

Scientists have discovered that rocks can be ejected into space in an impact event, and that it's possible that the rocks could protect small spores or amino acids in their interiors, which then could be transferred between planets. So it's even possible that life originated on Mars first, and then was transferred to Earth!

Some of this excitement about life on Mars comes from the discovery in 1996 of certain compounds in a Mars meteorite that seemed to be the result of life. Since we don't have any direct samples taken from the surface of Mars yet, scientists have to study the handful of Mars meteorites which have landed on Earth and been identified. One such meteorite, named ALH84001, was studied carefully, and scientists thought it contained a variety of chemical compounds and physical structures that appeared to be microscopic remains of biological activity. Perhaps the most exciting of these discoveries was the presence of so-called microfossils, structures in the rock that looked a lot like tiny, microscopic, fossilized organisms!

FACTS

New Mars missions are based on the premise, "Follow the water." Equipment on future orbiting and landing craft will search for carbonates (rocks that form in the presence of water), subsurface water, and signs of water (past or present) in rocks and soil on the surface.

Other scientists have studied the rock, however, and the evidence looks a lot less conclusive than it once did. Nevertheless, this discovery, followed by the landing of Mars *Pathfinder* on Mars in 1997, reignited public excitement about the possibility of life on Mars. When we also consider that there could be a current reservoir of liquid water beneath the surface of Mars, there is not only the possibility of fossilized life, but also perhaps current life surviving underground! (Refer to Appendix B for statistics on all the planets described here, and in Chapter 7.)

CHAPTER 7

The Outer Solar System

The inner solar system and outer solar system are divided by the asteroid belt. The outer solar system has four giant gaseous planets and Pluto, a small icy world at the far reaches of the solar system. The four gas giants have all been visited by one or more unmanned spacecraft, but Pluto has never been studied up-close at all.

Jupiter

FIGURE 7-1:
The planet
Jupiter

(refer to page
278 for more
information)

Courtesy of NASA/Jet Propulsion Laboratory (JPL), Caltech

Jupiter, appropriately named after the Roman king of the gods, is the largest planet in the solar system, and has more mass than all the other planets combined. Jupiter is one of the brightest objects in the sky, after the Sun, Moon, and Venus (and sometimes after Mars), and is easily visible at certain times of the year. Even a small pair of binoculars will show you up to four points of light lined up on either side of Jupiter. These are the four large satellites of Jupiter, called the *Galilean satellites* because of Galileo's fateful discovery that they orbited around Jupiter.

SSENTIALS

Galileo's observation of Jupiter's satellites proved that not all objects in the solar system orbited Earth, and helped provide support for the Sun-centered view of the solar system originally proposed by Copernicus.

If you look at Jupiter in a small telescope, you can begin to see bands, and sometimes the Great Red Spot. You can also track the progress of the four satellites (Io, Europa, Ganymede, and Callisto) as they orbit Jupiter and seem to switch places on either side of the huge planet. If you're lucky, you might see one of the satellites transit in front Jupiter, or go behind it, only to emerge later on the other side. Such careful observation helped Galileo determine that these really were satellites of Jupiter.

If you have a midsized or large telescope, Jupiter is one of the most satisfying solar system objects to look at in the night sky (perhaps after the Moon and Saturn). For instance, the scars due to the impact of comet Shoemaker-Levy 9 into Jupiter in 1994 were easily visible from the ground using a medium-sized telescope.

Jupiter is a gas giant. Unlike the rocky, terrestrial planets of the inner solar system, Jupiter is a giant atmosphere without a solid surface underneath! Actually, there might be a small rocky or metallic core at the center, but astronomers aren't sure yet whether or not this core exists. The central core would be about ten to fifteen Earth masses, puny when you consider that Jupiter's total mass is about 318 times that of Earth. Jupiter is mostly made up of the lightest element, hydrogen (90 percent), with a lesser amount of helium (almost 10 percent) and small amounts of methane, ammonia, water, and rocky or metallic material at the core.

QUESTIONS?

Where did planets get their names?
Planets are named after gods and goddesses from Greek and Roman mythology. Most of these gods have two names—one in the Roman language, one in Greek. Interestingly, Uranus is the only planet named after a Greek deity—all the rest are Roman.

From the outside, the most striking things about Jupiter are the stunning clouds. Compositionally, scientists believe that there are three separate layers of clouds made up of different mixtures of ammonia ice and water ice. From orbit, though, what you notice are the brightly colored bands, ranging in color from reds to yellows to whites. These colors are mostly due to the chemical composition of each layer, including small amounts of sulfur, which can appear in many different colors. The winds on Jupiter blow at very high speeds (more than 400 miles per hour), and the wind blows in opposite directions in bands that are next to each other. This wind system means that the bands rotate in different directions, and explains why moving pictures of Jupiter have a very odd appearance.

The light bands are called *zones*, and the dark ones are called *belts*. At the boundaries between bands are complicated hurricane-like storms that form due to the opposite rotational directions. Perhaps the most famous feature on Jupiter is the Great Red Spot (GRS), which has been observed from Earth for the last 300 years and could have been around even longer. The GRS is an oval-shaped storm that is twice as big as Earth. Scientists still don't understand how a storm that big can be so stable and last so long.

The energy that drives the winds and storms on Jupiter comes mostly from the heat released from inside Jupiter, not from the Sun. In fact, Jupiter releases more heat into space than it receives from the Sun. Remember that Jupiter is five times farther from the Sun than Earth is, meaning it gets much less sunlight. Jupiter also has a strong magnetic field, which traps lots of energetic particles near Jupiter. These high-energy particles result in a huge amount of radiation, causing problems for space probes like *Galileo*, and certainly threatening any astronauts who might try to explore Jupiter or its moons.

FACTS

Most of Jupiter is not made up of material in a normal gas state, but of a strange substance called *liquid metallic hydrogen*, which can form only under extremely high pressures. As far as we know, it is found only in the interiors of Jupiter and Saturn.

Jupiter's rings are very difficult to see from Earth, unlike the spectacular rings seen around Saturn. They were discovered by the *Voyager 1* spacecraft during its flyby in 1979. The rings are very dark, and are made up of small bits of rock and dust thrown off Jupiter's tiny inner moons by small meteor impacts. As this ring material falls into the planet due to drag from the atmosphere and magnetic field, the moons supply new material to the rings.

The Galilean Satellites of Jupiter

Jupiter has sixteen confirmed satellites, though recently observers have discovered new moons that have yet to be confirmed. The four largest moons, the Galilean satellites, are the most interesting. Io, Europa, Ganymede, and Callisto (all named after various lovers of the god Jupiter/Zeus) form what is almost a mini–solar system, with the surface age increasing and amount of geologic activity decreasing with distance from Jupiter. The inner three (Io, Europa, and Ganymede) are locked together in a tidal resonance—for every one time Ganymede orbits Jupiter, Europa orbits twice and Io orbits four times. This resonance keeps the orbits from being circular, resulting in varying tides. The height of the tide raised in each moon by Jupiter (like the tide raised in Earth by the Moon) varies during each orbit, thereby producing a flexing and pulling of the satellite that ends up heating the satellite's interior.

Io

In the case of Io, the innermost Galilean satellite, the evidence of tidal flexing is obvious. Io is the most volcanically active body in the solar system, and although it is only about the size of Earth's Moon, it is even more volcanically active than Earth! Current volcanic activity on Io was one of the most amazing discoveries found at Jupiter by the *Voyager* spacecraft. Compositionally, Io is a mostly rocky body with a large iron core, like Earth. However, the surface of Io is covered with various volcanoes that are constantly erupting in the form of plumes, explosions, lava flows, and other more exotic types of volcanic activity. This activity is so constant that no impact craters have yet been found on the surface of Io, meaning that it has one of the youngest surfaces in the solar system. Many obvious changes were found on Io between the two *Voyager* flybys, only four months apart in 1979 and 1980, and more changes were found when the *Galileo* spacecraft arrived fifteen years later in 1995.

FIGURE 7-2:
The four Galilean satellites

From left: Io, Europa, Ganymede, and Callisto

(refer to page 278 for more information)

Courtesy of NASA/JPL/Caltech

Some of the lava flows on Io are silicate flows similar to those found on Earth. The bright reds, oranges, and yellows seen on much of Io's surface led *Voyager* scientists to believe that most of the volcanic activity was due to sulfur, but measurements taken by the *Galileo* spacecraft have shown that the temperatures of many of the volcanoes are too high to be sulfur, and in some cases almost too high to be normal silicate lavas! Scientists are still puzzling over what exactly is going on at many of the volcanoes.

Europa

If Io is a world of fire, Europa, the next moon out, is a world of ice. Images from the *Voyager* and *Galileo* spacecraft revealed a satellite covered with an icy surface, broken in places by long, cracklike features that extend for hundreds and in some cases thousands of kilometers.

The early images of Europa were mostly low resolution, and showed only linear features and regions of fuzzy-looking mottled terrain. Images taken starting in 1996, however, revealed a much more interesting surface. The long linear features are actually huge ridge and crack systems that cover much of the surface. Scientists now believe that these ridges are formed as a result of tidal flexing.

FACTS

If most of its icy layer were liquid water, Europa could have more water than all Earth's oceans combined! And that's quite a lot— there are about 3.612×10^{20} gallons of sea water in our oceans.

Scientists believe that the flexing from Europa's tides causes the crust to crack and break in certain places. Once a crack is formed, it opens and shuts every few days as Europa orbits around Jupiter. This motion could force ice and slush up from beneath the surface, material that could then be squeezed out onto the surface and eventually build up the huge ridge systems that crisscross Europa's surface as if winding a ball of string. Another possibility is that warm material rises up to the surface at zones of weakness, like volcanic activity on Earth. The tidal heating of Europa could also produce something even more interesting—an ocean of liquid water below the icy surface! Scientists believe there might be enough tidal heating to keep water liquid below a surface cap of ice.

In addition to these models of tidal heating, there is other evidence for subsurface liquid water on Europa. Regions of disrupted terrain on the surface exist where blocks of the surface seem to have come apart, moved around, and refrozen in new positions. These ice blocks are similar to icebergs on Earth, only much larger. The best evidence so far for an ocean comes from magnetic field measurements consistent with a global salty water layer.

Scientists are very excited about the possibility of water below the icy surface of Europa because where there is water, there could be life! In fact, some believe that Europa is the second-best place to look for non-Earth life in the solar system, after Mars. The ice at the surface of Europa is probably too thick to let sunlight through to the water. Thick ice therefore rules out most photosynthesis, the method by which much simple life on Earth gets its energy. However, there could be volcanoes at the bottom of the ocean that give out heat, chemicals, and energy, and other necessary materials created at the surface by radiation processing could make their way down into the ocean. It is possible that life could survive there, much like it does at the hydrothermal vents at the bottom of Earth's oceans.

Ganymede

The next two large moons of Jupiter, Ganymede and Callisto, might also have subsurface oceans. If they exist, however, they are so deeply buried beneath the surface that they would be very difficult to reach. The surface of Ganymede is primarily covered with craters, but there are also smooth regions covered with ridges and grooves. These features are most likely due to some kind of tectonics, but just how they formed is still unknown. The areas of Ganymede with ridges and grooves are younger than the old, heavily cratered areas, but even the youngest parts of Ganymede are still quite old, much older than the surface of Europa.

Ganymede is the largest moon in the solar system. It's even larger than the planet Mercury, although Mercury has much more mass. Ganymede is also much larger than the planet Pluto, which is so far out that there the Sun appears as only a very bright star.

The craters on Ganymede are similar to craters elsewhere in the solar system, but are generally much flatter and shallower. These characteristics are seen in the few craters on Europa as well, and may exist because

Ganymede's surface is made up mostly of ice. Unlike rock, ice can flow over time like a glacier. Scientists think that with time, some of the craters on Ganymede have even relaxed away completely, leaving circular flat areas called *palimpsests*. Ganymede has its own internally generated magnetic field, meaning that below its icy surface there could be a metal core with a layer of rock on top of it.

Callisto

Callisto, the most distant of Jupiter's large moons, has the oldest surface. Its exterior layer is completely covered with impact craters of various sizes and ages, ranging from tiny craters up to huge basins that are surrounded by multiple rings. These features are thought to occur when a huge object hits the surface and causes the subsurface layers to flow, resulting in a wrinkling of the surface layer. The surface of Callisto seems to be covered with a thick dusty layer that might be shaken around by impacts, since most of the smallest craters seems to be filled in.

One of the great puzzles in studying Jupiter's moons is why Ganymede and Callisto are so different. They're about the same size, located in orbits right next to each other, and have similar compositions. Yet Ganymede is a fully developed world containing tectonic activity, an iron core, a magnetic field, and other features. Callisto, by comparison, is very primitive. Its interior is most likely a mixture of rock and ice all the way down, although current research suggests that it might be somewhat differentiated, and its surface is covered only by craters. One possibility is that since Ganymede is in resonance with Europa and Io, and Callisto is not, this discrepancy could result in just enough tidal heating of Ganymede to give it the energy needed for internal geologic activity, while Callisto stayed cold and inactive.

Saturn

Saturn, the ringed planet, is named after the Roman god of agriculture. It is a fairly bright object in the sky, usually located near Jupiter. Through binoculars you can sometimes see an oval shape depending on the orientation of the rings, and a small telescope can easily reveal the rings separated from the planet in all their glory. In fact, Saturn is probably one of the most satisfying targets for a telescope.

FIGURE 7-3:
The planet
Saturn

(refer to page 278 for more information)

Courtesy of NASA/The Hubble Heritage Team

Saturn's composition is similar to Jupiter's, approximately 75 percent hydrogen and 25 percent helium, with small amounts of water, methane, ammonia, and rocky materials. Its internal structure is similar to Jupiter's, as well—a small rocky core, then a weird liquid layer of metallic hydrogen, followed by a huge gaseous exterior. Saturn's atmosphere has dynamics similar to Jupiter's, but Jupiter's bands are not nearly as striking. Instead of a Great Red Spot, Saturn has a variety of white ovals, which are also long-lived storm centers.

Rings

The most stunning feature of Saturn is its ring system. The rings of Saturn are named with various letters. The two most prominent, the A and B rings, as well as the fainter C ring, are all visible from Earth. Images taken by the Voyager spacecraft showed four fainter, additional rings. Saturn's rings are different from the rings of the other giant planets—they are bright instead dark. These bright colors are the reason why the rings are visible from Earth, and are probably due to the rings' composition of water ice particles. The rings are not solid disks of material, but are made up of uncountable small particles, ranging in size from a centimeter to a few meters. There are also a few larger objects in the kilometer-size range.

FACTS

Although Saturn's rings are very broad—spanning over 250,000 kilometers out from Saturn, they are extremely thin—less than a kilometer deep. Saturn's tilt changes, which is why the rings seem to appear and disappear. When the rings are pointed edge-on at Earth, they are so thin that they aren't even visible!

The rings seem to be kept in place by small shepherding satellites, which orbit Saturn in the gaps between the rings or just inside or outside the smaller rings. The rings also have other odd structures, including spokes that seem to point out from Saturn like spokes on a wheel, and braids and knots seen in the smaller, outer rings. Like all rings around planets in the outer solar system, the lifetime of any particular ring particle is quite short, and the rings must constantly be supplied with fresh material ejected from the larger satellites.

Satellites

Saturn has eighteen currently named satellites, and a number of recently discovered ones that are still being verified. Of these, most are small, icy, cratered bodies. Titan, the largest satellite of Saturn, is an exception. Only a little smaller than Ganymede, Titan is the only satellite with a thick atmosphere. In fact, the atmospheric pressure at the surface of Titan is about 50 percent higher than that on Earth! The atmosphere is so thick and hazy that the *Voyager* spacecraft weren't able to take pictures of the surface. Similar to Earth's, Titan's atmosphere is mostly nitrogen, with a small percentage of argon, methane, and other organic compounds.

The surface below Titan's thick atmosphere is still mostly a mystery. It is possible that the interior of Titan is still hot, meaning that the surface could still be geologically active. Some clouds in Titan's atmosphere are made up of ethane and methane; these clouds could rain out onto the surface and produce lakes or perhaps even an "ocean" of liquid ethane and methane! Observations of Titan taken at infrared wavelengths by the Hubble Space Telescope, which can partially penetrate through the haze, have shown brightness variations on Titan's surface that could be continents in an ocean, though this theory has not yet been proved.

Uranus

The planet Uranus, the third largest, was named after the ancient Greek god of the Heavens. It is difficult to see from Earth and was discovered by William Herschel in 1781. Herschel first thought it was a comet before realizing it was a planet! Uranus is just barely visible to the naked eye when the sky is clear if you know just where to look—it's easier to find with binoculars or a small telescope.

Courtesy of NASA/Jet Propulsion Laboratory (JPL), Caltech

FIGURE 7-4:
The planet Uranus, stretched at right to show cloud layers

(refer to page 278 for more information)

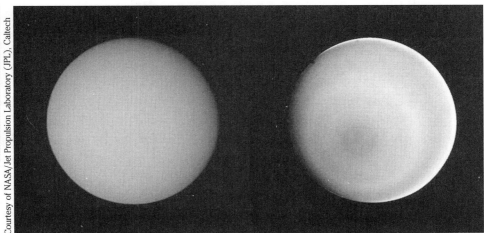

Uranus is an odd planet. Most of the planets in the solar system spin on an axis that's perpendicular to the plane of the solar system. However, Uranus spins on its side—when the *Voyager 2* spacecraft visited Uranus in 1986, the only spacecraft to do so, its south pole was pointing almost straight at the Sun! The magnetic field of Uranus is also odd. Most planets, like Earth, have a magnetic pole that's pretty close to the rotational pole. On Uranus, though, the magnetic field isn't even centered on the middle of the planet, and it seems to be tilted at a 60-degree angle with respect to the spin axis.

Uranus and Neptune have similar compositions, and are in fact quite different from Jupiter and Saturn. Uranus is made mostly of rock and a variety of ices, and only has about 15 percent hydrogen and a small amount of helium. Neither Uranus nor Neptune has the liquid metallic hydrogen that's found on Jupiter and Saturn.

Above its ice and rock surface, Uranus has an atmosphere composed mostly of hydrogen (83 percent), with 15 percent helium and about 2 percent methane. Uranus has bands of clouds and fast-blowing winds like the other giant gaseous planets, but there is very little color difference between the different cloud bands. During *Voyager 2's* visit, Uranus was perhaps the most boring-looking planet in the solar system, with a uniform blue color over almost the whole atmosphere. This blue comes from the methane layer in Uranus's atmosphere, which absorbs the red light from the Sun. If there are more brightly colored bands on Uranus, they are hidden below the uniform blue methane layer.

FACTS

A collision with an Earth-sized object millions of years ago could explain why Uranus has an asymmetrical axis and is tipped on its side, though scientists still are not sure of this or any other theory.

Although Uranus had a bland, boring appearance during *Voyager's* visit, more recent observations taken by the Hubble Space Telescope seem to show that Uranus is starting to change, and more streaks of color seem to be visible in its bands. It's possible that these changes are due to the changing seasons on Uranus—the Sun has gone from being over the south pole in 1986 to residing at a lower latitude, and will be located over the equator by the year 2007.

Uranus also has a ring system, like all the other gas giant planets. The rings of Uranus were first discovered in 1977 when astronomers watching a star pass behind Uranus (called an *occultation*) noticed that the star seemed to turn on and off as it passed behind six separate rings before it went behind the planet. Observers on Earth and the *Voyager* spacecraft have discovered more rings, for a total of eleven separate rings surrounding the planet. The rings are dark, as are those of Jupiter, and are made up of objects ranging in size from miniscule to a few meters in diameter. The outer, brightest ring is kept in place by two shepherding satellites, Cordelia and Ophelia.

The Satellites of Uranus

Uranus has twenty moons that have officially been named, plus at least one more that has not yet been fully confirmed. There are three classes of moons of Uranus: eleven small dark inner moons, all of which were found by the *Voyager 2* spacecraft; five large moons that were easily discovered from the ground; and then a series of more distant moons that have only recently been discovered. Uranus's moons are named after characters from works by Shakespeare and Pope, rather than mythology. The five large moons, Miranda, Ariel, Umbriel, Titania, and Oberon, are the most interesting of the Uranian satellites.

Miranda

Miranda, the innermost of the large moons, was well-observed by the *Voyager 2* spacecraft on its visit in 1986. It is perhaps the most interesting satellite of Uranus, and certainly one of the more bizarre moons of the solar system. When the first images of Miranda were visible, scientists didn't know what to think! Miranda's surface appears to be a jumbled mess, with huge cliffs, grooves, and valleys superimposed on normal-looking, old cratered areas. Some of the canyons are huge, with depths up to 20 kilometers (12 miles)! Some of the grooved areas change direction suddenly, producing an interesting chevron-shaped feature.

ESSENTIALS

All the large Uranian moons (Miranda, Ariel, Umbriel, Titania, and Oberon) are composed of ice and rock (approximately 40 percent ice and 60 percent rock).

At first scientists thought Miranda might have been cracked apart in a collision and then reassembled, like a jigsaw puzzle put together incorrectly. Up to five different disruption events could have taken place, and after each event the moon would have been reassembled from the remaining pieces, with parts of the inside sticking out in some areas.

Scientists have recently proposed a more likely theory: partially melted ices could have come up from the interior and covered portions of the surface. Lighter material would remain buried under heavier surface materials until it eventually worked its way up to the surface. However Miranda's jumbled surface was created, scientists were very surprised to find this much tectonic activity on Miranda, since Miranda is small and cold.

Ariel

Ariel, the next of the larger satellites, is the brightest moon of Uranus. Like all of the large moons of Uranus, its composition is basically half water ice and half rock. Ariel's surface is mostly covered with impact craters, but there are also large valleys and canyons. These valleys probably formed tectonically, as Ariel's crust stretched, and opened up faults on the surface. They could have formed in this manner if Ariel once had a liquid interior, but then froze solid and expanded (since water ice expands when it freezes).

Some of Ariel's valleys are hundreds of kilometers long, and more than 10 kilometers deep. Some of these valley floors are filled with smooth deposits that look as if a fluid material has flowed there. The temperature is much too cold on Ariel for it to have been water, but the flows could have been made by a more exotic material such as ammonia or methane.

Umbriel

The next Uranian moon, Umbriel, is the third largest. Umbriel's dark surface is covered with many craters, but bears no evidence of tectonic activity. There is a strange bright ring near the north pole which could be the floor of a large crater. The surface can also be divided into subtle dark and bright regions. The dark areas may be from an early period of icy volcanism, as seen on Miranda and Ariel.

Titania

Titania, the largest moon of Uranus, is mostly covered with craters. It does have evidence of faulting and other signs of tectonic activity, meaning that at some point in its history it was geologically active. One interesting feature is a valley that is 1,600 kilometers long! Titania actually looks very similar to Ariel, and one possibility is that the interior of Titania was once liquid, but froze, expanded, and caused the surface to crack.

Oberon

Oberon, the last of the large moons of Uranus, is similar in appearance to Umbriel. Its surface is covered with impact craters, and has more large craters than the younger surfaces of Ariel and Titania. Craters with rays of ejected matter have been found, and some crater floors are dark. The darker color could be due to upwelling material in the bottom of the crater. Oberon does have some faults, mostly near its southern hemisphere, which show that there was at least some geologic activity in the early part of its history. Since then, however, very little geologic activity has taken place on Oberon. Oberon also has a large mountain, which is 6 kilometers tall.

Neptune

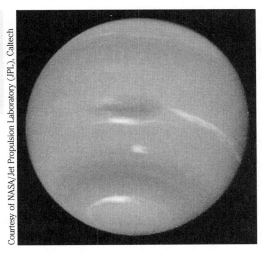

FIGURE 7-5:

The planet Neptune

(refer to page 278 for more information)

Courtesy of NASA/Jet Propulsion Laboratory (JPL), Caltech

Neptune, the smallest of the four gas giant planets, is named after the Roman god of the sea. It can be seen with binoculars if you have a very good locating chart, but you need a fairly large telescope to see anything more than a tiny disk. Observations of Neptune were recorded by Galileo in 1613 when it was near Jupiter, but he thought it was just a star (unfortunately, it was cloudy when he might have been able to track its motion). Neptune wasn't discovered as a planet until 1846. After the discovery of Uranus, scientists noticed that the orbit of Uranus wasn't quite in accordance with Newton's laws, and this observation required that there be another planet outside of Uranus influencing its orbit.

The Planet

The central portion of Neptune is made up of an Earth-sized rocky core, covered by a mixture of water, liquid ammonia, and methane.

Neptune's atmosphere is made mostly of hydrogen and helium, with small amounts of water and methane. As is the case for Uranus, the methane gives Neptune its blue color. However, the clouds have a deep blue color rather than the greenish blue of Uranus, meaning that there must be an unidentified chemical element coloring the clouds.

Neptune's atmosphere is very active. Neptune has the fastest winds in the solar system, reaching speeds of 2,000 kilometers per hour (1,250 miles per hour)! Like the other gas giants, it has large storms, and winds divided into different latitude zones. The only spacecraft to visit Neptune so far was the *Voyager 2* spacecraft in 1989. At that time, Neptune had a feature called the Great Dark Spot in its southern hemisphere. This feature was about the same size as Earth (about half the size of Jupiter's Great Red Spot). There were also a variety of smaller dark and white spots seen on Neptune, including one small white cloud that traveled all the way around Neptune once every sixteen hours! This feature, called the *scooter*, could result from a plume of material rising up from deeper within Neptune's cloud layers. *Voyager* also saw long bright clouds up high in Neptune's atmosphere, clouds that appeared similar to cirrus clouds on Earth.

SSENTIALS

The Great Dark Spot moved around Neptune at a speed of 300 meters per second (700 miles per hour) in 1989. Five years later, the Great Dark Spot was gone, but a new one appeared a few months later. These observations show that Neptune's atmosphere is quite dynamic and changes on short time scales.

Neptune also has a small ring system, with dark rings similar to those of Uranus and Jupiter. Four distinct rings have been observed— they were seen as only faint discontinuous arcs from Earth, but *Voyager* observations showed that they were complete rings with occasional bright clumps of material. One of the rings has an oddly twisted structure, like the braided rings seen on other giant planets. Neptune also has a magnetic field, which, like that of Uranus, is tilted from the axis of rotation and offset from the center of the planet. Scientists believe this odd orientation might be due to flows deep in the interior of the planet.

The Satellites

Neptune has eight currently known satellites. Seven are quite small but one, Triton, is fairly large and very interesting. Triton was discovered only a few weeks after the discovery of Neptune itself, in 1846. One odd fact about Triton is that it's in a retrograde orbit and therefore orbits Neptune backwards. A few small satellites of Jupiter and Saturn orbit backwards, but Triton is the only large moon in the solar system to orbit this way. Scientists believe this means that Triton couldn't possibly have formed in place around Neptune—it must have formed elsewhere in the solar system, and then been captured by Neptune.

Most satellites in the solar system have prograde orbits (they rotate in the same direction their planet rotates), and are slowly moving away from their planet. However, since Triton is in a retrograde orbit, it is slowly moving toward Neptune. Eventually, it will either crash into Neptune or simply break apart.

Triton's axis of rotation is tilted at a large angle with respect to Neptune's axis. Like the planet Uranus, Triton alternates between having its poles and its equator pointing directly at the Sun. When the *Voyager 2* spacecraft visited Triton in 1989, its south pole was pointing at the Sun, implying that Triton has very odd seasons as first one pole is in daylight, then the other.

The surface of Triton is quite different from most of the cold, dead satellites of the outer solar system. Its surface is quite young—there are very few impact craters visible on it. Most of the southern hemisphere of Triton is covered with a frozen ice layer of nitrogen and methane ice. The surface also has a complicated series of ridges and valleys that cover the surface. These features could result from varying cycles of freezing and thawing as different parts of Triton's surface move into sunlight. Triton also has smooth icy plains that could have been formed by volcanic eruptions of liquid water or a combination of water and ammonia.

One of the most exciting discoveries on Triton was active volcanic plumes! These ice volcanoes erupt plumes of material that were seen to rise 8 kilometers above the surface, and extend downwind over 100 kilometers. The erupted material is probably a mixture of liquid nitrogen, dust, and methane compounds. These eruptions are likely driven by seasonal heating from the Sun, as different parts of the surface are heated and then frozen again. Images of Triton taken by the Hubble Space Telescope in 1995 suggest that more of the south polar ice layer has evaporated since 1989, and that some of this material might have condensed in the equatorial regions creating a bright region there.

Pluto

FIGURE 7-6:
Pluto and its moon Charon

(refer to page 278 for more information)

Courtesy of NASA/JPL/Caltech

Pluto, the smallest planet in the solar system, is usually the farthest from the Sun (except when it crosses the orbit of Neptune). In fact, Pluto is smaller than seven of the solar system's moons! Pluto is visible from the ground with a midsized telescope, but it's very hard to spot. You'll need an excellent finding chart in order to have any chance of locating it.

Pluto was discovered in 1930, and named after the god of the underworld. Originally, scientists thought that the motions of Uranus and Neptune required there to be another planet out beyond the orbit of Neptune (similar to how Neptune was found in the first place). Clyde W. Tombaugh, working at Percival Lowell's Lowell Observatory in Arizona, made observations based on this information, and ended up finding Pluto. However, Pluto was much too small to account for the supposed motions of Uranus and Neptune, so astronomers continued looking for the so-called Planet X. No planet was found, and current measurements and calculations indicate that there is no tenth planet—and, in fact, Neptune's orbit does not require another planet beyond it.

The Planet

Pluto has a very strange orbit—it is highly eccentric, meaning that instead of having a path close to a circle, its orbit is long and stretched out. For twenty years during its 249-year orbit around the Sun, Pluto is actually closer to the Sun than Neptune is. Pluto just finished being the eighth planet from the Sun—this period lasted from 1979 to 1999. But Pluto has a high inclination, as well—its orbit is tilted by 17 degrees from the plane of the solar system (the ecliptic). So as Pluto crosses paths with Neptune, Pluto is far above or below the plane of Neptune's orbit, and there is no danger that they will crash into each other. Pluto is also locked in a 3:2 resonance with Neptune—for every one orbit that Pluto makes around the Sun, Neptune makes 1.5 orbits.

Pluto is made up of between half and three-quarters rock, with the rest various types of ice. Its icy surface is covered with mostly solid nitrogen ice, with smaller amounts of methane ice and carbon monoxide ice. When Pluto is at its closest approach to the Sun, a very thin atmosphere forms which then starts to freeze and fall back onto the surface as Pluto moves away from the Sun and its surface cools. The atmosphere, like the surface, is probably made up mostly of nitrogen with smaller amounts of carbon monoxide and methane.

Pluto is the only planet that has not been visited close-up by a spacecraft. Pictures taken from Hubble Space Telescope and other Earth-based telescopes can just barely see features on Pluto's surface, but astronomers have been able to map out bright and dark areas. In fact, Pluto is one of the most high-contrast bodies in the solar system—the bright areas are very bright, and the dark areas are very dark. Pluto has a very bright south polar ice cap, a dimmer north polar cap, and a variety of both bright and dark features in the mid-latitudes.

Pluto's Moon Charon

Pluto has one large moon, Charon, which is more than half as big as Pluto! Charon is the largest moon with respect to its primary planet in our solar system. Behind Pluto and Charon is our own Earth-Moon system. Because Charon is so close in size to Pluto, some scientists believe that Pluto and Charon are a double-planet system rather than a planet and a satellite. The density of Charon is lower than that of Pluto, though, meaning that it is made up mostly of various ices. Its surface is covered primarily with water ice, unlike Pluto's nitrogen surface. Overall, Charon's surface is much darker than Pluto's.

Pluto and Triton, Neptune's largest moon, are actually very similar, and both have very odd orbits. This similarity suggests to some scientists that there may be a link between them. At first they thought that Pluto might have been an escaped satellite of Neptune's, but that idea is now thought to be unlikely. The current prevailing theory is that Triton is a captured object, perhaps a remnant from the early days of the formation of the solar system, and that Pluto somehow escaped capture. Pluto, Charon, and Triton might be some of the only remnants from this time.

**SSENTIALS** NASA has been considering proposals for a Pluto mission for some time, but so far a mission is too expensive. There's considerable public support for a mission to Pluto, though, since it's the last unexplored planet.

Future Exploration of Pluto

Because so much is unknown, NASA is trying to launch a mission to study Pluto. There is some time pressure to perform such a study as expediently as possible, both for reasons of orbital mechanics (Pluto is easier to get to now than it will be in the future, due to planetary alignments) and because as Pluto moves farther from the Sun, its atmosphere will start to freeze onto the surface. Scientists would like to get a mission to Pluto within the next ten to fifteen years, since models of Pluto's atmosphere suggest that after that, most of Pluto's atmosphere will be frozen solid on the surface and difficult to study. Since the travel time to Pluto is at least five to seven years, a mission needs to be launched soon!

CHAPTER 8

Solar System Wanderers

Asteroids, comets, and meteors are different rocky and icy bodies that affect the solar system. Often grouped together under the heading small bodies, these objects are distinct classes with very different properties. Some objects, however, can transition from one class to another. Classifications are primarily defined by composition, speed, position in the solar system, orbit, and relation to Earth.

Asteroids

Asteroids are sometimes called *minor planets* because they resemble planets in many ways. But by definition, an asteroid is a rocky or metallic atmosphere-free body that orbits around the Sun, usually less than 1,000 kilometers in diameter. Most asteroids tend to congregate in what has come to be called the *main asteroid belt*—a ring-shaped area between the orbits of Jupiter and Mars. It's generally thought that asteroids are leftover remnants from the early stages of solar system formation that never got formed into a planet because of Jupiter's intense gravitational field.

QUESTIONS?

How are asteroids categorized?
Asteroids are commonly separated into types by two major factors: albedo and spectral composition. Albedo is the ratio of reflected light to incident (or source) light; white surfaces have an albedo of one, black surfaces have an albedo of zero. Spectral composition aids in identifying an object based upon the wavelengths of light the material absorbs or reflects.

The first asteroid known to modern times is named Ceres after the Roman grain goddess. Originally thought to be a new comet by its discoverer, Guiseppe Piazzi, in 1801, this mammoth asteroid measures about 600 miles in diameter. Today there are more than several hundred thousand named asteroids, though most are much smaller than Ceres, and a few thousand more are found each year.

Types of Asteroids

There are three basic compositional types of asteroids (plus a few other, much rarer classes):

C-TYPE (*carbonaceous*) are extremely dark due to their hydrocarbon content, and have a reflectivity of around 3 percent. C-type asteroids comprise the basis for some of the oldest bodies in the solar system.

Most asteroids, upwards of 75 percent, fall into the C-type category. Ceres and Pallas, for example, are both C-type asteroids.

S-TYPE (*silicaceous*) asteroids are bright because of their high iron composition. They are constituted entirely of silicate materials, and typically have a reflectivity of around 15 percent to 20 percent. These asteroids represent between 10 percent and 15 percent of the asteroid population.

M-TYPE (*metallic*) asteroids are comparatively brighter than S-type and C-type asteroids. They are composed of an iron and nickel alloy, and are often thought to be from the nuclei of other bodies. M-type asteroids, such as Psyche, are quite rare.

Asteroid Groupings

Asteroids can also be grouped by their solar system location:

MAIN BELT asteroids are located in the asteroid belt between the orbits of Mars and Jupiter. Most of the asteroids are in the Main Belt, and they can be divided into further orbital and compositional subgroups, each of which is named after the main asteroid in the group. The asteroids do not smoothly span the distance of the belt, however. Orbital resonances with Jupiter cause gaps in locations because any asteroid that happened to end up in that orbit would quickly be moved into a different orbit due to Jupiter's gravity.

The asteroid belt is an unstable place, and objects are often perturbed, or shifted, into orbits that cross paths with other asteroids. Impacts are common, and these impacts can send fragments of material into the inner solar system where they become meteoroids and can eventually impact Earth or other planets. Whole asteroids can also occasionally be perturbed into different orbits, making their way into the inner or outer solar system. The moons of Mars, and many of the small moons in the outer solar system, appear to be captured asteroids. These large objects occasionally impact the planets, producing spectacular impact craters.

NEAR-EARTH ASTEROIDS (*NEAs*) are asteroids that have been perturbed out of their main belt orbits and get especially close to Earth. Like main belt asteroids, NEAs are also grouped into various classes based on their orbits. Earth-crossing asteroids are those whose orbits cross the orbit of Earth. Scientists are especially interested in finding all the known NEAs because of the possibility of one hitting Earth; such an event could have catastrophic consequences. The Moon bears the scars of such impacts. The Near-Earth Asteroid Tracking (NEAT) program at NASA identifies new NEAs and provides carefully calculated orbits, producing up-to-date information on what we might expect in the near future.

TROJANS are located at stable points in Jupiter's orbit called *Lagrange points.* Several hundred of these objects have been located and as many as a thousand may exist in total. A few small asteroids may also be located in the stable Lagrange points of the orbits of Venus, Earth, and Mars.

The outer solar system also contains one strange class of objects, called *Centaurs*, that bear some similarities to asteroids, but probably contain a much higher percentage of frozen volatile gases and are better classified as comets. The most famous Centaur is Chiron, which orbits between Saturn and Uranus. Chiron was originally identified as an asteroid, but was recently reclassified as a comet. These objects are mostly in unstable orbits that are subject to change.

Famous Asteroids

Ceres carries great significance as the first documented asteroid. A mathematical formulation from the mid-1700s known as Bode's law predicted the finding of a planet between Mars and Jupiter, and the discovery of Ceres was originally thought to fulfill this prediction (until its small size was realized). Interestingly enough, Bode's law was also successful in predicting the location of Uranus. Ceres is large enough to be seen with the naked eye when its orbit brings it close enough to Earth. Heinrich Wilhelm Olbers, who identified further examples of minor planets, continued Guiseppe Piazzi's work, and asteroid hunting became an accepted part of astronomy.

The asteroid Vesta is one of the only other asteroids that can consistently be seen without binoculars or a telescope. It was observed through the Hubble Space Telescope between November 28 and December 1, 1994, and at that time the asteroid was only 156 million miles from Earth. Vesta is about 326 miles in diameter and has distinguishable light and dark areas, much like the craters on the Moon.

The chances of an asteroid or comet hitting Earth are small but not nonexistent. After all, a giant asteroid did kill off all the dinosaurs 65 million years ago, right? Scientists are always on the lookout.

Vesta is thought to be one of the few asteroids to be fully differentiated, meaning that it is sufficiently large and has undergone enough heating to separate into layers based on density, with a metallic core and a mantle made up of lighter, rocky materials.

Studying Asteroids

Scientists have been able to learn about asteroids in several ways. First, and perhaps most exciting, is the study of actual bits of asteroids that have fallen to Earth in the form of meteorites. When scientists examine these asteroid fragments firsthand, they gain insight into their chemical composition and internal structure. Telescopic observations allow us to study asteroids from Earth, and imagery from spacecraft such as *Galileo* and the *Near-Earth Asteroid Rendezvous (NEAR)* allows us to capture high-resolution digital images and compositional data for further study.

Scientists are currently attempting to match up meteorite samples to telescopic observations of various asteroids, based mostly on compositional data. Unfortunately, telescopic observations can measure only the composition of an asteroid's surface. The surfaces often have different chemical properties from the interior, so it's rarely certain which asteroid is the parent body for which group of meteorites.

Spacecraft Observations

The *Galileo* spacecraft flew past the asteroids Ida and Gaspra on the way to its main mission at Jupiter. These were the first close-up studies of asteroids, and revealed small, primitive bodies that had been scarred by millions of years of impacts.

The *NEAR (Near-Earth Asteroid Rendezvous)* spacecraft was a very interesting project led by Johns Hopkins University. Launched from Earth February 17, 1996, *NEAR* was scheduled to go into orbit around the asteroid Eros in January 1999. Eros was selected because it is one of the larger asteroids, and its orbit was most compatible with *NEAR*'s launch. Upon *NEAR*'s approach to Eros, however, the spacecraft didn't enter the orbit correctly and missed the asteroid. Luckily, the engineers were able to recover the spacecraft and send it around the Sun on a catch-up orbit in February 2000. After taking thousands of valuable pictures, *NEAR* actually landed on the surface of Eros in 2001, the first time a spacecraft had landed on an asteroid.

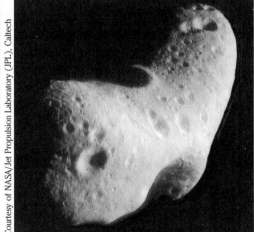

FIGURE 8-1:
The northern hemisphere of Eros

(refer to page 279 for more information)

Courtesy of NASA/Jet Propulsion Laboratory (JPL), Caltech

NEAR carried several specific instruments designed to study different aspects of Eros. A magnetometer engaged in a fruitless search for a magnetic field, while an x-ray/gamma-ray spectrometer weighed and measured chemical compounds. The multispectral imager documented the asteroid's shape and surface appearance, while the laser rangefinder was able to image the shape more precisely. *NEAR* helped scientists gain insight into the properties of this small, primitive body.

Close Encounters

Will an asteroid crash into Earth anytime soon? Some are forever predicting our untimely demise due to an asteroid or comet collision, but unfortunately these folks will likely have to wait awhile longer. There are a variety of programs surveying the skies to detect potential threats. In addition to NEAT, there is a program at the University of Arizona called *Spacewatch*. This program runs a 1.8-meter and a 0.9-meter telescope on top of Kitt Peak, an observatory in southern Arizona. Spacewatch is part of the Lunar and Planetary Lab at the University of Arizona and studies asteroids and comets, including those that might present a potential danger to Earth.

Comets

Because bright comets are easily observed, their history and discovery dates back far before asteroids, with some comets recorded in the third century B.C. There are roughly 1,000 comets that have been identified to date, with more discovered each year. It's fairly unusual to actually see a comet with the naked eye. A number of small comets are found each year, but most are detectable only with binoculars or a telescope. Small naked-eye comets come around approximately once every five or six years, but the ones with the huge tails, that can be seen even from major cities, are visible only every eleven or twelve years.

Defining a Comet

While asteroids are metallic, rocky bodies, comets are composed mainly of ice and nonvolatile dust. They can be made up of frozen water or gas, and are commonly called *dirty snowballs* or *dirty icebergs* because of their similarity to a time-worn snowman. These cold, icy bodies become active only when their orbits take them near the Sun. The Sun's heat causes the frozen gases to sublimate, and stream out from the comet's surface in vaporous jets. The jets are what can create a spectacular tail streaming out away from the Sun.

Active comets have several components:

Nucleus: a solid core composed of ice, gases, and dust
Hydrogen cloud: large, encompassing hydrogen area
Coma: dense material cloud of water, carbon dioxide, and gases
Dust tail: visible dust particles that escaped the nucleus
Ion tail: plasma particles that are generally caused by solar winds

When a comet's nucleus is still frozen, it is not visible directly; it can be seen only through reflected sunlight. Once a coma starts developing, the gasses will absorb ultraviolet radiation and begin to self-illuminate. When a comet gets close to the Sun, the nucleus heats up and gasses escape, causing them to stream large particle trails that allow us to observe them easily.

FIGURE 8-2:
The nucleus
of comet
Halley

(refer to page
279 for more
information)

Courtesy of NASA/NSSDC, Giotto

The dust tail is the main part of the comet that we see when looking to the sky. It can trail for up to 10 million kilometers and, depending on its position and other atmospheric conditions, can be quite visible. Sometimes both the dust and ion tails are visible, as in the case of comet Hale-Bopp. A comet's tail always points away from the Sun so that in some parts of the comet's orbit, the comet actually travels tail first!

Comet Travel

Comets are most readily seen when they are near the Sun. They have elliptical orbits that bring them close to the Sun before they swing out to where we cannot see them. A typical comet's orbit might last up to hundreds of thousands of years, so you won't necessarily see the same comet twice.

Comets can be divided into two classes based on the lengths of their orbits. Comets with long-period orbits can often sail out past Pluto into the Oort cloud; such long orbits, known as *eccentric orbits*, will bring the comet into a visible field only to have it disappear for thousands of years.

Short-period comets that come from the Kuiper Belt, such as Halley's comet, can stay inside Pluto's orbit for a much longer period of time. The Kuiper Belt is a zone of material extending out beyond the orbit of Pluto. Pluto, in fact, is sometimes called the King of the Kuiper Belt because it is the largest and closest body there. Since the Kuiper Belt is closer to the Sun than the Oort cloud, comets coming from there don't take as long to get near the Sun, and their orbits are shorter. The Kuiper Belt is also more closely confined to the plane of the solar system, so short period comets have smaller inclinations than long period comets.

QUESTIONS?

What is the Oort cloud?
The Oort cloud, named after Dutch astronomer Jan Oort, is a cloud of primitive material surrounding the solar system at great distances. The Oort cloud also extends above and below the solar system's plane, so long-period comets have large inclinations, and dip above and below the plane of the solar system on their trip around the Sun.

There are also a few other smaller classes of comets, including Jupiter-family comets, that have been captured by Jupiter's gravity. Perhaps the most famous Jupiter-family comet is comet Shoemaker-Levy 9, which broke into pieces on its second-to-last pass by Jupiter, and eventually crashed into Jupiter in 1994.

When a comet is near the Sun it might be observable with the naked eye for up to a few weeks, depending on the comet's orbit, Earth's atmospheric conditions, and other factors. The comet will likely be visible with binoculars and telescopes for a longer period of time while it is farther from the Sun and much fainter.

Famous Comets

Comet Halley was initially discovered by Edmund Halley, a British scientist living and working in the eighteenth century. He used Newton's laws of motion to determine that the comet observed in 1531, 1607, and 1682 would continue its average orbit of seventy-six years and return in 1758— which, in fact, it did. Unfortunately, Halley died in 1742 and never got to see his comet return.

Not all comets become bright! It is very difficult to predict just how bright a comet will become as it approaches the Sun. Some comets that are predicted to be the next comet of the century end up being unimpressive, while other spectacular comets appear to come out of nowhere.

Comet Hale-Bopp burst onto the scene in July of 1995 with a blaze of glory and ions. It made its closest pass by the Sun on April 1, 1997. Hale-Bopp was brighter than everything in the night sky, except for the Moon and the very brightest stars. It was discovered outside Jupiter's orbit by Alan Hale and Thomas Bopp, and was determined through the Hubble Space Telescope to be about 40 meters in diameter. Comet Hale-Bopp had an unusual double tail—both the dust tail and the ion tail were clearly visible from Earth.

Discovered by Yuji Hyakutake almost exactly a year later in 1996, comet Hyakutake was particularly exciting because it passed so close to Earth—within 9.3 million miles! (For comparison, the distance between Earth and the Moon is only 240,000 miles.) With its exceptionally long tail, Hyakutake could easily be seen with the naked eye, making it especially appealing to astronomers worldwide.

Comet Shoemaker-Levy 9 presented one of the most exciting comet discoveries in recent history. Between July 16 and July 22, 1994, parts of Shoemaker-Levy impacted Jupiter at speeds of 60 kilometers, thus creating the first observed series of collisions between solar system bodies.

Falling Stars

Meteors, meteoroids, and meteorites are different terms that can often be confused, but shouldn't be. Rocks or dust floating freely in space are meteoroids. A meteor is a shooting star in the sky—it's a space rock burning up in our atmosphere. If a piece of this material makes it all the way through the atmosphere and hits the ground, we call it a meteorite.

Meteors

FIGURE 8-3:
An artistic rendering of a meteor storm

(refer to page 279 for more information)

© 2001, www.arttoday.com

Have you ever seen a shooting star? If you've been out stargazing, you've probably noticed what looks like a star streak across the sky, then disappear. Shooting stars, or falling stars, are technically called *meteors*. Most are small grains of dust, or occasionally tiny pebbles, and are often leftover debris from comets. These are, in fact, the smallest particles that orbit the Sun. Shooting stars might appear blue and white, or perhaps yellow and orange—the color can depend on the speed at which the meteor is traveling.

Why do they look so bright? These tiny pebbles are little pieces of space junk that occasionally crash into Earth's atmosphere, but since they're so small, they usually don't make it to the ground. They enter the atmosphere at such high speeds that bits of gas trapped in the rock are vaporized, producing a very bright glow—the bright streaks we call *shooting stars*. Meteors are visible to us on Earth only when they pass into Earth's atmosphere.

Meteor Showers

Generally, one or two meteors will be visible on just about any night you're out looking at the stars. A few times a year, though, meteor showers occur. The most famous is the Perseids, which is visible every year around August 11 and 12. When Earth passes through part of its orbit that contains more dust and small rocks than usual, we see meteor showers.

Since we go around the Sun once a year, we pass through this dirty area once a year, predictably about the same time each year. Why is this particular area of space so dirty? Blame a comet! The reason it's dirty is because Earth crosses through the trail of debris left by a comet as it neared the Sun. The gas and dust blown off a comet as it partially disintegrates leaves a trail behind it, even after the comet itself is long gone.

FIGURE 8-4:
A fireball

(refer to page 279 for more information)

© 2001, www.artoday.com

The Perseids are exciting because on a good year, you can see as many as 400 meteors an hour, instead of only one or two as on normal nights. During a meteor shower, you're also more likely to see large bright meteors, sometimes called *fireballs*. The Perseids are the most reliable meteor shower.

Meteorites

In order to make it all the way down to the ground, a meteor has to be a solid rock, rather than an icy particle left over from a comet. Therefore, most meteorites that hit Earth actually come from the asteroid belt. As asteroids crash into each other, they break off bits of material that get sent toward the Sun. Some of these objects eventually intersect Earth's orbit, and fewer still manage to survive the fiery trip through the atmosphere to land on Earth's surface. Most meteorites are small enough to be significantly slowed down by their trip as their exterior layers burn up; what's left of them hits the ground (sometimes creating a crater on impact).

Scientists study meteorites to determine their chemical compositions and what physical processes they've undergone. Researchers also attempt to match up particular meteorites to individual asteroids. Since meteorites are one of the few samples we have of nonterrestrial material, scientists are always excited to discover a new meteorite to gain additional information on the solar system.

Impact Craters

Very occasionally, a very large piece of rock comes through Earth's atmosphere and hits the ground, making a crater. fifty thousand years ago, a large metallic rock about 1,000 feet across crashed into the desert in Arizona, and created Meteor Crater, which is almost a mile across. The explosion scattered bits of rock and metal up to 5 miles away from the crater. Similarly, craters on the Moon and other rocky bodies in the solar system are caused by comets or asteroids impacting with the planet's surface.

FIGURE 8-5:
Meteor Crater in Arizona

(refer to page 279 for more information)

© 2001, www.arttoday.com

Asteroid and Comet Impact: Dinosaurs

Scientists believe that about 65 million years ago, a giant comet or asteroid hit Earth, resulting in global disaster. The impact would have blown enough dust and gas into the atmosphere to effectively block the sunlight and make Earth much cooler, ultimately changing its climate. Cold-blooded dinosaurs would be unable to adapt to the change in climate, and mammals would have the opportunity to begin their period of domination.

ESSENTIALS

The giant impact and the resulting change in plant and animal life would disrupt the entire food chain from plants, to herbivores, to their predators. The lesson, here, is that without mammals' ability to adapt to Earth's changing ecosystem, we might not be here today.

For years, scientists wondered what could have caused the mass extinction that is visible in the fossil record of 65 million years ago. About fifteen years ago, some scientists began to notice that there is a layer of iridium located in rocks dating from the time of the extinction. Iridium is rare on Earth, but is common in meteorites. Scientists also found tiny grains of shocked quartz in this layer, indicating that they had been in an extraordinarily forceful event.

Although these two findings suggested that an impact event had occurred, there was no evidence of the event itself. The missing piece was eventually located when a large, underwater impact crater was identified off the Yucatan Peninsula in Mexico. The Chicxulub crater was found to be 65 million years old, which corresponds to the giant impact event that caused the dinosaurs' extinction.

The giant impact theory of extinction had a profound effect on the study of geology and evolution. Originally, scientists had assumed that change was slow and steady over geologic time. Small changes occurred over eons, in the same way a river slowly carves a path through a valley, eventually producing significant changes over millions of years. The impact theory, however, showed that change could sometimes occur due to one huge, spectacular event and its effect on both the appearance of the landscape and the environment.

CHAPTER 9

A Star Is Born

Everyone is familiar with the life cycle here on Earth—birth, aging, and death. To us, stars seem to be static, permanent objects, but they go through a similar cycle. On time scales of millions to billions of years, stars change dramatically, going through different stages until their own eventual death. Depending on its mass, a star's aging process can lead to anything from a white dwarf to a supernova.

Stellar Structure

Stars are primarily composed of the simplest, lightest, and most common element in the universe: hydrogen. For example, our own star, the Sun, is made up of about 94 percent hydrogen, 6 percent helium, and very small amounts of heavier elements. Stellar energy begins in a star's core, through nuclear fusion reactions. Nuclear fusion requires very high temperatures and densities. The Sun's core is about 16 million degrees Kelvin, and is about twenty times more dense than iron. It's not a solid rock or metallic core as the planets' are; solar temperatures are too high for the materials there to be anything except gas.

Beyond the Core

Around the Sun's core is a radiative zone, where energy travels from the interior to the exterior layers. In the radiative zone, energy travels through the radiation of particles of light (photons). The outer 15 percent of the Sun is called the *convective zone*; in this cooler outer layer, the gasses' circular motion (convection), moves heat and energy.

The surface of the Sun is called the *photosphere*, which is the deepest part of the Sun that you can actually see. Here, photons escape the Sun and begin their journey to Earth or wherever else they might be headed. The surface of the Sun, at a mere 5,840 degrees Kelvin, is where sunspots are located. Sunspots are regions on the surface that are cooler than the surrounding areas, and that appear to move across the face of the Sun as it rotates.

Sunspots and Solar Flares

The Sun has an extremely strong magnetic field, and often ejects bursts of material in solar flares. These bursts of high-energy radiation can make it to Earth and beyond, and sometimes disrupt satellite communications here on Earth. They also cause bright auroras in the northern and southern skies when they interact with Earth's own magnetic field. Solar flares often erupt from particularly prominent sunspot groups.

Galileo used sunspots to map the rotation rate of the Sun. Of course, since the Sun is not a solid body, different parts of it rotate at different

FIGURE 9-1:
The Sun

(refer to page 279 for more information)

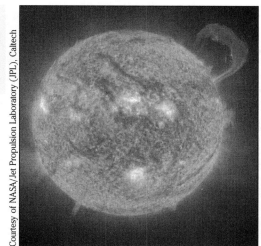

Courtesy of NASA/Jet Propulsion Laboratory (JPL), Caltech

speeds (the same phenomenon occurs on giant gas planets like Jupiter). The equator of the Sun takes about twenty-five days to rotate, while regions at a latitude of about 60 degrees take closer to thirty days to rotate.

The safest way for an amateur observer to see sunspots is from a projection of the Sun. You can create such a projection by pointing the telescope at the Sun (remember, don't look through the finder scope) and placing a piece of white cardboard a few feet away from the eyepiece. (You may have to adjust the telescope and cardboard's location to get the image to focus.) To create a projection with a pair of binoculars, place a piece of white cardboard behind the eyepieces of the binoculars (which should be on a tripod), and move the binoculars until you get a white disk on the cardboard. The dark spots you see on the projection of the Sun are sunspots.

ALERT

You can do permanent damage to your eyes by looking straight at the Sun—whether you are observing with the naked eye, or through a telescope or binoculars. Buy a special solar filter for your telescope or binoculars. These items are usually used during solar eclipses, but can be used anytime.

Stellar Lives

The main factor that determines what course a star's life will take is its mass. Remember that stars shine because nuclear fusion reactions are performed in their cores, where less dense elements, such as hydrogen, are fused into more dense elements, such as helium. A star's total mass determines how much fuel it has to burn, and these reactions are what give off heat and light.

Large stars have more fuel to sustain themselves than smaller stars do, but a large star is also brighter, so it burns its fuel faster. This rate of expenditure gives large stars a shorter overall lifetime. Our own Sun is on the small side, and will burn for about 10 billion years before it uses up all its hydrogen fuel. By comparison, a star that has sixty times the mass of our Sun will have a very short lifespan of only 3 million years, while a star that's only 0.1 times (or 1 percent) the mass of the Sun will last for many billion years.

Star Formation

Let's take a look at the typical life cycle of a star. As we discussed, stars condense from gaseous material. The first stars formed soon after the Big Bang, but star formation continues even today. Stars are born in giant molecular clouds, or nebulae, which are huge clouds of gas and dust that exist between stars—in our galaxy and others.

FIGURE 9-2:
A nebula

(refer to page 279 for more information)

Courtesy of NASA/The Hubble Heritage Team (STScI/AURA)

Regions of the molecular cloud begin to condense and stick together, and eventually gain enough mass to ignite and start producing light. One molecular cloud can be the birthplace for dozens of stars. The Orion Nebula, for instance, has produced four very bright stars, and more are in the process of forming. Since a number of stars can form from one molecular cloud, they initially appear in clusters, locations where a number of young stars of the same age exist very close to each other.

The key to gasses and matter condensing is the force of gravity, which causes all matter to attract other matter. Initially in a nebula, one small area might be slightly denser (and therefore have slightly more mass) than another area. The mass in the denser area will attract other mass, eventually forming a small ball that continues to collect mass from the surrounding nebula in a process called *accretion*. Once the object collects enough mass, it begins to collapse, again under the force of gravity, and the star becomes much hotter.

As a gas clump collapses, the gas particles get closer and closer together and bump into each other more, generating heat. The initial clump in the gas cloud forms a protostar, an object that gives off infrared and microwave radiation, but is not yet performing nuclear fusion. The clump eventually forms a disk with the protostar in the center, and the material surrounding the protostar in the disk will eventually either form more stars, or turn into a planetary system like our own.

Mature Stars

Once a protostar collapses enough and becomes sufficiently dense, fusion begins in the core of the star. The pressure from the fusion reactions keeps the star from continuing to collapse. After fusion starts, the star ejects most of the remaining dust and gas surrounding it by producing strong winds. This phase is called T-Tauri after the first star observed doing this. The material is thrown away from the star in huge jets of gas and dust. Once the star settles down and stabilizes, a main sequence star is left. During this stable phase, the star fuses hydrogen atoms together to form helium, and gives off heat and light in the process. Over time, the original cluster loses many of its stars due to gravitational forces from other stars; single stars (known as *field stars*) are the ones that have escaped along the way.

Since mature stars lose their companions quite early, if you see a cluster of stars together in the sky, they are probably young. One observable star cluster is the Pleiades. They are visible in the constellation Taurus, and are shaped like a very small version of the Little Dipper. The Pleiades, much younger than our Sun, are only about 80 million years old.

Red Giants

A star spends most of its lifetime in a stable, main-sequence phase. This phase must come to an end when the star has used up all its fuel, and converted all its hydrogen into helium. When the nuclear reactions in the star's core stop, the star starts to collapse, again, due to an imbalance between gravity and the support that came from internal energy. This collapse causes the gas to compress and heat up, and fusion begins in the layer outside the core. As the star continues to collapse, heat increases and speeds up fusion—much like the cycles that occurred during the protostar phase. The increased heat and activity in the core makes the star brighter again, and the layer of hydrogen gas surrounding the star, the envelope, starts to expand and cool and the star becomes first a subgiant, then a red giant.

A red giant is basically a star with a bloated outer surface. Since the star's energy is spread out over a larger area as its surface expands, the surface itself becomes cooler and has a redder color, hence the name red giant. In our solar system, when the Sun becomes a red giant star a few billion years from now, it will expand beyond the orbits of Mercury and Venus. Although Earth most likely will not be eaten up by the Sun, Earth's water and atmosphere will be boiled off, making it uninhabitable by today's standards.

QUESTIONS?

Do planetary nebulae have anything to do with planets?
Early astronomers thought these formations looked like planets, and were so named. In fact, planetary nebulae are much larger than planets—most are even larger than our solar system!

Size Matters

Once a star reaches the red giant phase, its mass becomes important. If the star is much more massive than our Sun, it will continue to expand and become a supergiant. Betelgeuse, located in the constellation Orion, is a supergiant. Betelgeuse is so big that it would fill our solar system out to the orbit of Jupiter!

Some stars, if they are massive enough, will go through core fusion, at which point they start fusing helium and eventually heavier elements. When this new fuel runs out, the star collapses again, turns back into a red giant or supergiant, then cycles through core fusion and collapse repeatedly. The number of times a star can cycle through these steps depends on its mass— the greater its mass, the more frequent its cycles. Each time the star goes through a fusion cycle, it creates heavier and heavier elements through nucleosynthesis. Small stars like our Sun can generate only enough heat to create lighter elements such as carbon and oxygen.

Eventually, a star will run out of fuel. If it was a low-mass star to begin with (up to five times the mass of our Sun), then it will eject its outer layers as the core shrinks down. This ejected material forms a planetary nebula, and the leftover remnant of the star's core at the center is called a *white dwarf star*. One example of a planetary nebula is the Ring Nebula (see Chapter 10). Others, like the Dumbbell Nebula and the Cat Eye Nebula, have more asymmetrical and complex shapes due to different formation processes.

FIGURE 9-3:
The Ant
Nebula, a
planetary
nebula

(refer to page
279 for more
information)

Courtesy of NASA and The Hubble Heritage Team (STScI/AURA)

Supernovas and White Dwarf Stars

As you'll remember, there are a few parts to the life cycle. Birth, growth, and—eventually—death. A star's mass will determine its ultimate fate. A large, massive star will come to a flashy, spectacular finish, while its more modest counterpart may simply collapse, cool off, and fade away. Then again, it may not . . .

Supernovas

When a high-mass star (five to fifty times the mass of our Sun) runs out of fuel, it goes through these reactions before it explodes into a Type II supernova:

1. The dense iron core of the star implodes, producing neutrons and neutrinos.
2. The core collapses incredibly quickly and turns into a neutron star.
3. Energy from the collapse helps heat the gas in the outer layers until it becomes superheated.
4. An explosion begins, and elements heavier than iron, such as gold and uranium, form. The superheated gas explodes out into space, carrying a variety of heavy elements out with it.

The final explosion is called a *supernova*. As the supernova expands more and more into the surrounding space, it loses its energy and eventually becomes a supernova remnant.

Because there aren't that many high-mass stars in the universe, supernovas are quite rare—in a given galaxy, one supernova will form every 100 years. But supernovas are extremely bright, so even if none has happened near us recently, we can still see them as bright objects in the sky. The bright phase of a supernova or planetary nebula only lasts for a few thousand—or tens of thousands—years (a mere moment, in astronomical time). As the giant cloud of gas and dust expands and cools, it also gets dimmer. The material of the nebula eventually dissipates and spreads into the interstellar medium, which is the very low-density region between stars. Once the gaseous nebula is gone, all that's left is a core remnant.

Stellar remnants are very odd objects made of gas that is super-compressed by gravitational forces to make what's called *degenerate matter*. Depending on the star's size, one of these things can happen:

- **It will form a white dwarf.** If the star is small, leaving a central core less than 1.4 times the mass of our Sun, then the core turns into a white dwarf star. A typical white dwarf is about the size of Earth.

- **It will form a neutron star.** If the central core is between 1.4 and 3 times the mass of our Sun, it forms a neutron star. A neutron star is even smaller than Earth—usually only about the size of a city!
- **It will become a black hole.** If the central core is more than 3 times the mass of the Sun, it will keep on collapsing. Gravity dominates, and a super-compact object called a *black hole* is created (see Chapter 11).

Like anything in nature, stars follow rules. But also like anything in nature, any variance in circumstances will produce different results. Remember that in a star's life cycle, everything has to do with mass, so changes to its mass for any reason will affect it in the end.

White Dwarf Stars

When our Sun's red giant days are over, it will one day turn into a white dwarf star. A white dwarf will typically collapse until it is about the size of Earth. Then the pressure from the electrons in its interior will regain balance with gravity, and it won't collapse any more. Even so, a white dwarf is extremely dense. The only star denser is a neutron star, which is made from a star with a core between 1.4 and 3 times the mass of our Sun (see Chapter 11).

Since white dwarfs form after a star has used up all its fuel, they don't burn or generate any energy on their own. The light they give off is simply from the star cooling and radiating away excess heat.

Normally, a white dwarf will just slowly cool off and fade away into oblivion. However, sometimes white dwarfs are in binary star systems, in orbit with or near a more active star. Sometimes gas thrown off of the companion star will land on the surface of the white dwarf. The gas (mostly hydrogen) gets heated on the white dwarf's surface, and eventually gets hot and dense enough for nuclear reactions to start up again! Because these reactions are taking place on the surface of the white dwarf, not in the core, the results are very different.

When nuclear reactions begin in this case, hydrogen is thrown off the surface, creating a bright shell of hot gas that expands to surround the white dwarf. To observers on Earth, it looks like a new bright star has suddenly appeared! These objects were called *novas* by early astronomers because *nova* means *new* in Latin. Although the nova quickly fades away, more gas from the companion star can build up on the star's surface, and another nova burst can take place.

Observing Stars

Remember, all the stars you can see in the sky are part of our own galaxy. As you look up at the sky, though, you can see that some stars are brighter than others. In some cases this apparent distinction is simply because the brighter stars are closer, but in other cases, the stars have a higher luminosity, and give off more light.

The brightness of a star as seen from Earth is called its *magnitude*. Originally, stars fell into six different magnitude classes, with the brightest stars as first magnitude and the dimmest stars visible to the naked eye as sixth magnitude. In the 1800s, though, astronomers managed to measure the actual brightness of stars—not just how they were perceived by the naked eye—and determined that a first-magnitude star is 100 times brighter than one of sixth magnitude. The refined magnitude system, therefore, states that a difference of five magnitudes (six minus one) signifies a difference of 100 times in brightness.

 **SSENTIALS** Luminosity is the amount of light put out by a star. Stars that are more massive give out more light, so to an observer they will look brighter than a smaller star that is the same distance from Earth.

Some stars have magnitudes of 0 or even negative numbers, meaning that they're extremely bright! For example, the Sun's magnitude is about –26.7. The planet Venus, at its brightest, has a magnitude of about –4.4. The brightest star, Sirius, has a magnitude of about –1.4, and the bright star Vega has a magnitude right around 0.0. The faintest star you can see with your naked eye is about 6.5, and the faintest detectable objects in the sky (with highly specialized equipment) have a magnitude of about 27 or 28.

Remember that stars with higher masses can put out more energy, and therefore have a greater luminosity. A dim star that is close to us can appear brighter from Earth than a high-luminosity star that's much farther from us. We can distinguish between the two with *apparent magnitude* and *absolute magnitude*. Apparent magnitude is the brightness of the star as seen from Earth. Absolute magnitude is the brightness of the star as measured from a standard, consistent distance (defined as ten parsecs). In a simple, regimented universe, all the stars might be exactly the same distance away, and we could see their absolute magnitude. The brightest stars would also be giving off the most energy. Our galaxy is hardly regimented, however, so keep in mind that a star can be bright because it is giving off a lot of energy, or because it's close to us, or both.

Binary Stars

Some stars are in binary systems, meaning that two stars orbit each other. Often, one of the stars is much larger than the other, and in this case the smaller star seems to orbit the larger one. In fact, they're both orbiting their common center of mass (a point between the two stars whose location depends on the mass of each star). Binary star systems are useful because astronomers can measure the periods of each of the star's orbits, and determine the masses of the objects using some basic laws of physics.

If a binary star system is pointed edge-on toward Earth, then it is known as an *eclipsing binary* because one star will periodically travel in front of the other star. Since almost all stars are so far away that they just appear as tiny points of light in a telescope, astronomers can use eclipsing binary stars to figure out the diameters of the two stars. They monitor how much light is coming from the star, and when one star passes in front of the other, the amount of light decreases. The length of time the star's brightness is decreased is used to figure out the diameters of the two stars.

Other Solar Systems

Given the number of stars in the sky, there must be other solar systems and other planets out there, right? You would think so, but finding extrasolar planets turns out to be much more difficult than one might imagine. First, remember how stars and planets are created. A star forms by gravitational attraction of particles in a big cloud of gas and dust. As the cloud condenses and forms a protostar, the star begins to spin. The rest of the leftover gas and dust eventually falls into the equatorial plane of the star and forms a disk. Most of the material from this disk falls into the star, but some stays in orbit long enough to form planets, accreting out of the leftover dust in a manner similar to how the star formed in the first place.

Home

In our solar system, we have four dense rocky inner planets, then four giant gaseous outer planets, then several small, icy, leftover bodies farther out (of which Pluto is the largest). Scientists have to explain how our solar system came to be in order to understand how other solar systems can form. They believe that in the regions of the disk of leftover material near the Sun, the temperatures were too high for volatile materials like hydrogen and water to condense. Only rocky materials could become solid there, so the inner four planets are made mostly of rock. The outer solar system, starting with Jupiter, falls on the other side of the snow line—the distance from the Sun where it was cool enough for materials such as gases and water from the cloud to condense. The outer planets are, therefore, gaseous. This model also explains why the satellites of the outer solar system are mostly icy instead of rocky.

How to Find Faraway Planets

Scientists believe that other solar systems, systems around other stars, should form in a similar manner. But since stars are so bright, and planets are small and dim, finding them around other stars turns out to be quite difficult. In fact, a typical planet would be about a billion times fainter than its star! Even the most sensitive telescopes on Earth can't detect something that faint because the light from the bright star would drown it out.

Astronomers had to think of different ways to look for signs of other planets. One of the first clues was the presence of dust and gas disks surrounding some young stars. Disks surround a number of bright stars, including Vega, Beta Pictoris, and Fomalhaut, so these stars could very well be in the process of forming planets around themselves!

But what about stars that already have planets? Remember that with binary star systems, the two stars actually orbit their common center of mass. If one star is much larger than the other, the small star will orbit while the larger star will just seem to wobble in place. This same phenomenon occurs with planets, but the star wobbles even less. Still, scientists can detect this wobble if the planet is big enough and close enough to its star. Scientists look at the star's spectrum, and watch its spectral lines shift back and forth, indicating the star's motion.

ALERT

Binary stars and double stars do differ. Although double stars appear to be close to each other, they can actually be far apart. Binary stars, however, actually orbit each other in their own system.

Another way to find smaller planets is by using the *transit method*, which is similar to the eclipsing binary scenario discussed earlier. Remember that eclipsing binaries are binary star systems that are tipped edge-on, so that one star appears to pass in front of the other, causing its brightness to dim temporarily. If there are any planets orbiting a star, the planets will pass in front of the star, again causing its brightness to dim temporarily. If these variations in brightness could be detected, then astronomers would have another way to detect planets. At this point, however, no planets have been found using the transit method.

Future Studies

In the future, NASA and the European Space Agency both plan to launch spacecraft to look for planets surrounding other stars. NASA's *Kepler* Mission will look for planets using the transit method. The *Kepler* spacecraft hopes to find planets that are, at the smallest, a few times the size of Earth. It won't be able to find planets as small as Earth.

CHAPTER 10

Deep-Sky Objects

In their quest to identify comets, early astronomers also found faint, fuzzy objects that resembled comets, but were not. They compiled detailed catalogs of these nebulae, which initially just referred to faint, fuzzy objects. Nebulae later referred to objects consisting of large gas and dust clouds. Other noncomet objects that contained stars and were farther away than typical nebulae, were termed galaxies.

Object Classification

When early astronomers looked up into the sky, they saw several classes of objects. Stars, arranged into constellations, were one of the primary classes.

A second class of objects was then identified—objects that moved against the background of the constellations. These objects included planets and comets, and were special because they could be tracked. As positions of planets and patterns of comets could eventually be predicted, it became simpler to classify these known objects separately from other things in the sky.

A third class of objects included ones that looked fuzzy, both to the naked eye and through early telescopes. These objects were not comets because they did not move; they stayed in the same place at all times relative to the background of the constellations. They also didn't resemble pinpoints of light, like stars. These objects were termed *nebulae*, and astronomers created detailed catalogs of their positions to distinguish them from comets. French astronomer Charles Messier (1730–1817) was the first to compile such a catalog, so some of these objects came to be called *Messier objects*.

FACTS

Deep-sky objects are defined as any objects outside of our own galaxy, including other galaxies, nebulae, globular clusters, and galactic clusters.

During the nineteenth century, the term nebula encompassed all fuzzy, nonmoving, nonstar objects in the sky; essentially, anything that wasn't a star, planet, or comet was a nebula. Later, the distinction between galaxies and nebulae came about as the result of research by Edwin Hubble (1889–1953) and others.

In the early twentieth century, Hubble used redshifts to show that nebulae were much farther away than any stars in Earth's sky, and his work showed that when galaxies move away from Earth, the speed at which they travel is proportional to the distance at which they are located. For example, Hubble showed that what scientists thought was the Andromeda Nebula was actually the Andromeda Galaxy—an entirely separate galaxy a number of light-years from our own. Astronomers computed distances for all the known nebulae, and concluded that many objects once considered odd, gassy regions in the sky were actually galaxies of their own!

Based on this new research, the nebula class became more specific: extragalactic nebulae are beyond our galaxy, and galactic nebulae are within it. Over time, of course, the general definition of nebula has changed. In current astronomical terms, a nebula is a large cloud consisting of dust and gas that exists in interstellar space, the space between stars. Many nebulae are part of our own galaxy.

A galaxy, on the other hand, is a large, organized structure that is very far away. Galaxies are groups of stars, gasses, and dust held together by gravitation and separated from similar systems by vast regions of space. Our own solar system is part of the Milky Way galaxy, as are all the individual stars we can see in the sky. Although most galaxies just look like faint fuzzy objects from here, they may contain large numbers of stars and planets, and perhaps whole civilizations!

Galaxies

There are several major types of galaxies, including spiral, elliptical, lenticular (convex on both sides), and irregular. Shapes vary, from disk galaxies (which lie mostly in one plane, like spiral and lenticular galaxies) to galaxies that are fully three-dimensional. They also vary greatly in size. The shape and structure of galaxies are governed by gravity, as well as by the initial conditions under which they formed. The distinctions between various types of galaxies are sometimes unclear, however, and some galaxies seem to include characteristics of multiple galaxy types. For example, some spiral galaxies have an elliptical central bulge, and some elliptical galaxies have a flat disk-shaped region.

Spiral Galaxies

Spiral galaxies come in two types—regular spiral (S) and barred spiral (SB). They are mostly flat structures that lie predominantly in one plane. Spiral galaxies have multiple arms of stars that appear to spiral out from a central core. The objects inside these arms are all in orbit around the center of the galaxy. Barred spiral galaxies have arms of stars that spiral out from a linear bar instead of a core. Spiral galaxies tend to have lots of dust and gas, material that can make it more difficult for us to observe them.

FIGURE 10-1:
Whirlpool
Galaxy M-51,
a typical
spiral galaxy

(refer to page
279 for more
information)

Courtesy of NASA and The Hubble Heritage Team (STScI/AURA)

The best example of a spiral galaxy is our own! The Milky Way galaxy has about a hundred billion stars, and the Sun is one of them. If you can observe from a dark place—one without a lot of ambient light at night—you should see a band of bright stars stretching from horizon to horizon. Ancients called this band the Milky Way, and it is actually the plane of our galaxy.

When you look up and see the Milky Way, you are looking toward the center of our own galaxy, where there is a higher density of stars. The center is obscured by many layers of gas and dust, so it's not visible to the naked eye, but infrared equipment allows astronomers to view our galaxy's core. Our solar system and Earth are located on one of the spiral arms of the Milky Way, meaning that we're fairly far from the densely populated galactic center.

QUESTIONS?

What's the difference between a galaxy and a nebula?
Galaxies are organized groups of stars, gasses, and dust. Nebulae are large clouds consisting of dust and gas. Nebulae are vague both in reality and definition, giving rise to the term "nebulous."

Spiral galaxies are one of the most beautiful sights in the sky, and astronomers have wondered how they got their shapes ever since they were first observed through telescopes. The current theory is that they actually formed that way. Remember that stars form from denser regions in an initial nebula of gas and dust. Early in their histories, some nebulae may have gravitational interactions with other nebulae or neighboring galaxies. These encounters cause clumps in the cloud of gas and dust, often in a spiral pattern. When stars eventually form, they retain the spiral pattern from the original nebula.

Elliptical Galaxies

The majority of galaxies fall into the elliptical category. Some elliptical galaxies, such as M87 in the center of the Virgo cluster, can have masses reaching up to a million times the mass of the Sun. Very large indeed! M87 is, in fact, a supergiant elliptical galaxy. Other elliptical galaxies, however, have more modest sizes, and many do not contain very much interstellar matter. Through a small telescope, they look like smooth, egg-shaped fuzzy bright patches.

Lenticular Galaxies

The less common lenticular galaxy is disk-shaped, but does not have much structure in its disk. The lack of structure in the disks could be because lenticular galaxies have used up most of their interstellar matter and are made up predominantly of old stars. It could result from the lack of encounters with other galaxies, thereby preserving their smooth appearance. Due to their oblong, convex shape, lenticular galaxies appear quite similar to elliptical galaxies, and are often misclassified as such.

Irregular Galaxies

A fourth type of galaxy, the irregular galaxy, contains stars that are not arranged in any orderly pattern. Some of these galaxies contain stars arranged with no observable symmetry, and some are distorted in such a way that their irregularity may be the effect of a nearby galaxy. M82, for example, is an irregular galaxy that falls into the distorted disk category. Its distortion is the result of gravitational interactions with its neighbor, M81. Other irregular galaxies include the Magellanic Clouds, which are small satellite galaxies of our own Milky Way and are visible only from the Southern Hemisphere.

FIGURE 10-2:
Galaxy ESO 510-G13, a twisted, warped galaxy

(refer to page 279 for more information)

Courtesy of NASA and The Hubble Heritage Team (STScI/AURA)

Active Galaxies and Galaxy Clusters

Active galaxies are not defined by their physical shape or structure. These galaxies have a core, or nucleus (the active galactic nucleus), which is thought to generate huge amounts of energy. Active galaxies emit this energy at various wavelengths, including radio waves. Black holes are theorized to be at the centers of such galaxies (see Chapter 11).

Galaxies can exist alone or in clusters (groups of galaxies that orbit near each other). The Virgo Cluster is one of the most visible examples. Clusters can contain hundreds or even thousands of galaxies, and sometimes spiral or elliptical galaxies make up the centers. Galaxies that are clustered together should, by the laws of physics, require a certain mass to be able to stay together. Observations suggest, however, that although the groupings have not achieved this specific mass, they seem to be stable; this dilemma is known as the *missing mass problem.*

Star Clusters

Stars can also be grouped into clusters smaller than multibillion-star galaxies. There are two major types of star clusters that can form within a galaxy: globular clusters and open (galactic) clusters.

FIGURE 10-3:
Globular
Cluster
NGC-6093

(refer to page 279 for more information)

Courtesy of NASA and The Hubble Heritage Team (STScI/AURA)

Globular clusters are very old clusters of stars with little structure. Containing anywhere from ten thousand to a million stars, their size can vary greatly. These dense groupings are spherical in shape. Ranging in size from 50 to 300 light-years across, they are unrestricted by the galactic plane, meaning they were formed before matter started settling into one plane. All the stars in a globular cluster were created at the same time, so they are all the same age, but the component stars can be different sizes, temperatures, and even colors. In general, stars in a globular cluster are all very old and have a low abundance of heavy elements, since they have not gone through multiple supernova cycles that would have enriched the surrounding medium with heavier elements.

Open, or galactic, clusters are groupings of stars that are less dense than a globular cluster. They have an irregular layout, usually with neither a central core, nor more than a few thousand stars. Open clusters were formed after the creation of the galactic plane, and as such, they are restricted by it. In some younger open clusters, star formation is still in progress. Since they are smaller than globular clusters, open clusters are subject to gravitational interactions as they travel through their orbits around the galaxy—these interactions can actually pull stars from the cluster. The average lifetime of an open cluster is only a few hundred million years. One well-known example of an open cluster is the Pleiades (M45).

Messier's Catalogue

Cataloging systems came about because of the need to track individual objects for research. Charles Messier, a French astronomer who compiled an initial list of over 100 objects, first became interested in astronomy upon his sighting of an enormous comet in 1744. From then on, he worked in various fields relating to space science. He observed an Andromeda companion galaxy in 1757, then discovered both a new comet and a nebula in 1758. This nebula, later named the Crab Nebula, would be the first object in his catalog.

Discovering comets was one of the newest, most exciting things do to in the eighteenth century, and the goal of Messier's catalog was to create a list of all visible objects that were clearly *not* comets. Eventually, the catalog became one of the most important sources for the locations and attributes of deep-sky objects. Objects in the *Messier Catalogue* are designated with an *M* as the first letter, followed by a number. The first edition of the catalog was published in 1774, including objects M1 through M45.

Galaxies seem to have halos of globular clusters surrounding them. Our own galaxy has a concentration of globular clusters—138 have been identified—around the galactic center. Some larger galaxies, such as M87, may have halos of up to several thousand clusters surrounding them!

Over the years, Messier continued to find and document objects in the sky—usually while searching for comets. In the process, Messier learned many important things about the sky, including that Uranus was indeed a planet and not a comet. His final list had 110 Messier objects. These 110 objects are some of the best deep-sky objects to observe with a small telescope. (Refer to Appendix C for a list of the 110 Messier objects.)

New General Catalog

The *Messier Catalogue* is not the only catalog of nonstar, deep-sky objects. In 1888 Johann Dreyer (1852–1926) published the *New General Catalogue of Nebulae and Cluster Stars*. After working at various observatories, Dreyer settled at the Armaugh Observatory in 1882. He obtained a large government grant, purchased a ten-inch refractor telescope, and set to work documenting the skies.

Items identified in this catalog all have new general catalog (NGC) numbers, which are different from the numbers of Messier objects. Some deep-sky galaxies, nebulae, and other objects have both Messier and NGC numbers; the NGC is much more inclusive, containing 7,840 identified objects. Updates to the NGC were published in the index catalog (IC), so some objects have IC numbers. The NGC catalog can be purchased as a book or in a computer-ready format, and can also be downloaded from the Internet.

Visible Galaxies

Two of the many galaxies visible to us in the northern sky are the Andromeda galaxy and the Triangulum galaxy. The Magellanic Clouds, satellite galaxies of the Milky Way, are visible only from the Southern Hemisphere.

Andromeda Galaxy

The Andromeda galaxy (Messier 31, NGC 224) is the closest large galaxy to us. The Andromeda system also contains a number of companions, including two bright dwarf ellipsoid galaxies. Andromeda is visible with the naked eye. It was first described by Al-Sufi in his *Book of Fixed Stars* from the first century A.D., and has been observed by astronomers ever since.

FACTS

Andromeda's radial velocity was measured at the Lowell Observatory in 1912. Moving at about 300 kilometers per second, it was the fastest-traveling object ever recorded at that time. Since radial velocity is correlated with distance, this discovery gave an indication of Andromeda's distance from Earth.

In 1923, Edwin Hubble continued the research into Andromeda, eventually determining it to be a separate galaxy instead of a nebula. Hubble had a lifelong interest in space and all things far away. After becoming a lawyer in the early 1900s, Hubble switched professions to become a full-time astronomer. Using Hooker Telescope at the Mount Palomar Observatory, Hubble saw stars all around Andromeda, proving definitively that it was a galaxy.

The Andromeda galaxy is one of the most-observed galaxies today, partly because it's so easy to observe. It has a number of companion galaxies, or neighboring galaxies, that are sometimes assimilated. M32 and M110 are two of the most prominent and the most visible, even with regular binoculars. M32 is a bright, small elliptical galaxy that can be readily observed when viewing Andromeda. It is about 3 billion solar masses in size, quite small compared to Andromeda itself. M110 is Andromeda's other bright companion, and is a small elliptical galaxy. A picture of Andromeda and two of its companion galaxies is visible in Chapter 3.

Triangulum Galaxy

The Triangulum galaxy, initially discovered by Italian astronomer Giovanni Hodierna (1597–1660), is also visible in the northern sky. Denoted in the Messier system as M33, Triangulum was one of the first-known spiral galaxies. M33 is about 3 million light-years away, and often the most distant object you can observe with the naked eye on a very clear night. Its area is about four times that of a full Moon, so it definitely takes up a large part of the night sky. M33 has at least five identified, bright globular clusters, although it may be surrounded by as many as twenty.

Satellite Galaxies of the Milky Way

The Magellanic Clouds, visible from the Southern Hemisphere, are some of the closest galaxies to us, and actually consist of two galaxies—the Large Magellanic Cloud (LMC) and the Small Magellanic Cloud (SMC). The LMC was first discovered, as the name suggests, by Magellan in the early sixteenth century. The SMC was discovered at the same time but is slightly farther away from us than the LMC. The SMC orbits the Milky Way at a distance of about 210,000 light-years, where the LMC's orbit is about 179,000 light-years.

The closest external galaxy to the Milky Way is the Sagittarius Dwarf Elliptical galaxy, just discovered in 1994 by Ibata, Irwin, and Gilmore. It is located only 80,000 light-years away, making it by far the closest galaxy. Part of the Local Group, Sagittarius Dwarf and its companions are associated with the Sagittarius constellation, and there are also neighboring globular clusters such as M54.

Nebulae

Once observers were able to distinguish between galaxies and nebulae, they could look at the sky and all that lies beyond it in whole new way.

Galaxies are exciting because they are of another place. Nebulae are just as exciting, however, because they are literally in our system's back yard.

Types of Nebulae

Nebulae are still well-studied today. These clouds of gas and dust provide different types of information about the past and present of the universe, as well as their own formation. There are several different types of nebulae:

REFLECTION NEBULAE: Dust and gas clouds that reflect the light from nearby stars. They tend to be on sites where stars formed years before. Because they reflect light, they usually appear blue. NGC7023 is an example.

EMISSION NEBULAE: Very hot gas clouds. Light from nearby stars is absorbed and re-emitted at a lower energy level. The balance of the absorbed energy becomes radiation. Since hydrogen is the predominant element in these types of energy exchanges, emission nebulae tend to appear reddish. M42 is one example. Current star formation often takes place in emission nebulae.

PLANETARY NEBULAE: Formed from the gas thrown out when a small-sized star (like our Sun) is approaching the end of its lifetime. The term planetary nebula is actually a misnomer; nebulae resemble planets when seen through a small telescope, but have nothing else in common with them. M57, the Ring Nebula, is one example.

DARK NEBULAE: Clouds of gas and dust with one specific property: they are large and dense enough to block the light of an object behind them. The Horsehead Nebula and NGC2264 are examples.

SUPERNOVA REMNANTS: Created at the end of a large star's life, when it undergoes a huge explosion and sends most of its mass into space. Discarded mass creates a supernova remnant, a slowly expanding cloud of gas and dust. M1, the Crab Nebula, is one example of a supernova remnant.

Crab and Ring Nebulae

FIGURE 10-4:

The Ring Nebula, a planetary nebula

(refer to page 279 for more information)

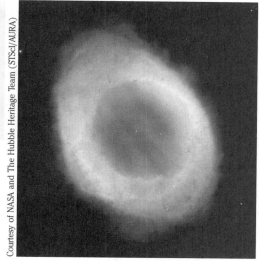

Courtesy of NASA and The Hubble Heritage Team (STScI/AURA)

The Crab Nebula (M1, NGC 1952) is located in the southern area of the constellation Taurus. Messier discovered this supernova remnant in 1758 as he was looking for comets. This nebula was one of the first objects he observed to be not a comet and not a planet, and it was designated M1. The supernova that created this nebula exploded in A.D. 1054, and was observed by Imperial Chinese astronomers. The Crab Nebula is 6,500 light-years away from us.

The Ring Nebula (M57, NGC 6720), a favorite target for mid-sized telescopes, has also been captured photographically by the Hubble Space Telescope. The colors of this planetary nebula range from blue in the center to orange and red on the edges; radiation from a remnant white dwarf star in the middle causes the gas to glow in different colors. This nebula is about 2,000 light-years from Earth, and measures about 1 light-year in diameter.

FIGURE 10-5:

The Horsehead Nebula

(refer to page 279 for more information)

Courtesy of NASA and The Hubble Heritage Team (STScI/AURA)

Eagle and Omega Nebulae, and the Pleiades

The Eagle Nebula (NGC 6611), an emission nebula, appears as several giant columns of dust and gas particles, mainly hydrogen. The name comes from an apparent visual similarity to the talons of an eagle. Young stars near the nebula give its gasses the same type of glowing effect seen in the Ring and other nebulae. The Eagle Nebula is about 2 million years old, and is located approximately 7,000 light-years from Earth.

The Omega Nebula (NGC 6618, M17), another emission nebula, is in the constellation Sagittarius. Other names for it include the Swan Nebula, the Horseshoe Nebula, and the Lobster Nebula. Although located close to the Eagle Nebula, it is much brighter, and therefore more visible. M17 is in an area where young stars were formed but the stars themselves cannot be seen, indicating that they are either inside the nebula, or are very young. M17 is mainly reddish from hydrogen gas; the central area is whitish, probably due to a combination of the gasses and reflected starlight. This nebula is between 5,000 and 6,000 light-years away.

FACTS

The *Local Group* of galaxies is a term used to describe objects close to us. These objects include the Milky Way, the Magellanic Clouds, the Andromeda Galaxy, the Triangulum Galaxy, and their companions.

The Pleiades (M45) is an open star cluster, containing six or more bright stars surrounded by a cloud of gas and dust. In clear weather, at least twelve stars can be observed with the naked eye. There are more than 500 stars in this grouping but they are very spread out. First observed as early as 1,000 B.C., this star cluster contains the Pleiades constellation, or the seven sisters, which is shaped like a small dipper. The blue color of the surrounding gas cloud helps identify this region as a reflection nebula, meaning that it reflects the brightness of the stars near and inside the nebula.

CHAPTER 11

Faraway and Far-Out

The universe contains a variety of strange and exotic objects: huge star-sized objects that spin in periods of seconds, black holes so dense that even light cannot escape, and galaxies that give off radio signals that can be received here on Earth. Now, quasars and pulsars, wormholes and white holes could all be the next big thing to astronomers worldwide.

Origins of Black Holes

What do black holes look like? No one knows for certain, since it is impossible to actually see a black hole. Usually, objects in space give off light or other types of electromagnetic radiation, but black holes are so dense that nothing escapes them, not even light. We have no instruments for seeing black holes, so all of our knowledge in this area is theoretical.

Laws of Gravitation

To get the general idea of black holes, you just need to know this: Einstein's theory of relativity basically states that gravity is a curvature of the space-time continuum. Any body of matter attracts any other body of matter. The more matter there is, the stronger the pull.

For relatively weak curvature created by smaller-sized masses, Isaac Newton's law of universal gravitation applies. As the story might have gone, Newton observed an apple falling from a tree; he noticed that the apple accelerated as it fell, and understood the nature of the gravitational force acting on the apple. Newton generalized these ideas, and discovered that the Moon orbits Earth because of Earth's gravitational pull. Any objects in space exert gravitational attraction toward each other, hence the law of universal gravitation.

FACTS

The maximum speed at which anything can travel in the universe is the speed of light, 186,000 miles per hour. The speed of light can be called the universal speed limit, since it is the fastest any object can travel.

Objects that are either very big or very dense exert a stronger gravitational force than lighter ones do. As an object gets larger and larger, it becomes so large and dense that the force exerted by its gravity is greater than anything else around it. Such objects with incredibly strong gravity are called *black holes*.

How Do We Know They Exist?

One way to understand black holes is to look at the concept of escape velocity, the speed at which something needs to travel in order to escape the gravitational pull of a star or planet. For example, if astronomers want to launch a rocket away from Earth, the rocket needs to be going at a certain speed in order to escape Earth's gravitational pull, and not simply go into orbit around Earth or fall back to the surface.

FIGURE 11-1:
Jet of electrons emitted from galaxy with black hole at center

(refer to page 279 for more information)

Courtesy of NASA and The Hubble Heritage Team (STScI/AURA)

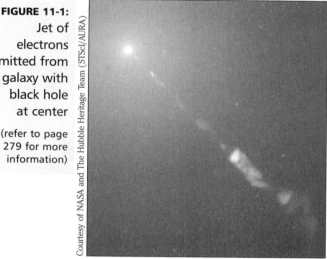

As the mass of the star or planet increases, the velocity required to escape it also increases. Therefore, sufficiently large objects reach a size where the escape velocity is the speed of light. This condition presents the threshold for a black hole: if there exists a situation where the escape velocity of a body must be greater than the speed of light, nothing can escape from it—not even light. Such objects are what we call *black holes*.

Where They Come From

The currently accepted theory as to the origin of black holes is that they form from collapsed stars. Remember that smaller stars can become white dwarfs or neutron stars. Stars on the upper end of the mass spectrum—with initial masses of up to fifty times the mass of our Sun—will come to a different end. If the central core of a collapsed star ends up with more than three times the mass of our Sun, it overcomes all forms of internal pressure and keeps on collapsing until it forms a black hole!

A black hole's properties depend not only on its large mass, but also on its very small radius. In fact, a black hole could theoretically have just about any amount of mass, as long as it was compressed into a small enough area (and therefore had a high enough density). However, scientists believe that black holes form only from the collapse of large stars, so they would also have large masses.

What *Is* a Black Hole?

Most of what science knows about black holes is theoretical, but certain structural elements are universal. Based upon these structural elements, scientists have determined ways of classifying black holes into different types based on certain other properties.

Singularity and Event Horizon

Black holes consist of two parts: the singularity, and the event horizon. The singularity is described as the central core of the black hole, where the radius of the black hole is equal to zero. A singularity can be described as a single point.

The event horizon is the distance from the center of a black hole, inside which everything approaching the black hole falls inward toward the singularity. In other words, anything that passes the event horizon would get sucked into the center (singularity) of the black hole with no hope of escape. The event horizon of a typical black hole is shaped like a sphere, and it can be thought of as the location where the escape velocity is equal to the speed of light. The distance between the event horizon and the singularity is called the Schwarzschild radius, and is usually considered the radius of the black hole itself. A black hole's *Schwarzschild radius* is proportional to the amount of mass the black hole contains.

Types of Black Holes

After a black hole forms, it can take one of three possible forms: Schwarzschild static black hole, Kerr rotating black hole, or Reissner-Nordstrøm charged black hole. It has also been theorized that the centers of many galaxies, including our own, could contain high-mass black holes, but this idea has not yet been proven. These super-massive black holes would have very different properties and formation processes from the stellar-mass black holes considered in this chapter.

Karl Schwarzschild (1873–1916) was a German astrophysicist who first calculated the radius of a black hole's event horizon. A black hole with a mass ten times the Sun's would have a radius of about 30 kilometers. Schwarzschild black holes are defined as those with a rotation of zero.

QUESTIONS?

Could Earth ever get sucked into a black hole?
No, because the Sun is much too small to ever become a black hole. Even if it did, we are too far from the Sun for the gravitational pull to drag us in.

Kerr black holes rotate, unlike static Schwarzschild black holes. Roy Kerr, a mathematician from New Zealand who studied rotating stars in the 1960s, hypothesized that since stars rotate, black holes probably rotate also. Kerr black holes can have ring-shaped singularities. Spinning black holes also have what's called a *static limit*. The static limit has a radius larger than the event horizon, and is sometimes called the *edge*, or *outer boundary*, of the black hole.

Once you cross the static limit of a rotating black hole, you are subject to its effects, which in this case means that it is impossible to stay still—everything inevitably gets dragged along in the direction of rotation of the black hole. Since the static limit is outside the event horizon, however, crossing this boundary is not final. It is possible to enter and leave the zone between the static limit and the event horizon at will, and an object or light ray only gets trapped forever once it crosses the event horizon.

Reissner-Nordstrøm nonrotating black holes are similar to static ones, but they are charged and have two event horizons. This type of black hole is thought to have an electric charge (called Q in physics equations), but it would likely attract the opposite charge as well, balancing it out. The outer event horizon is called the *static limit*, similar to the outer edge of a rotating black hole.

Appearance

What might black holes look like? First, they would be black because no light can escape from them. Around the outside of the event horizon, you might see light from stars behind the black hole, ones that were too far away to get sucked into the event horizon.

Black holes are one of a few types of celestial objects that cannot be seen, either with the naked eye or with advanced equipment. Rather, we know of the existence of black holes tangentially—large masses left over from supernovas, for example, could indicate a black hole. Similarly, astronomers can infer the mass of a central object by measuring the speed at which other objects orbit it. A large number of rapidly orbiting celestial bodies could suggest the presence of a black hole.

Stellar-mass black holes can be detected if they are part of a binary system—if one star of a binary star pair turns into a black hole, it can pull material away from the remaining companion star into an accretion disk of material surrounding the black hole. As the material in this accretion disk is pulled into the black hole, it gives off x-ray radiation that can be detected here on Earth. Thus, scientists believe that many x-ray binary systems contain a black hole.

Black holes actually exist in four dimensions: X, Y, Z (the three coordinates), and time. However, time inside a black hole is thought to be stretched out. Einstein's theories of time as a fourth dimension allow scientists to understand that a body on the edge of the event horizon has no choice but to fall into the singularity.

The fact that we cannot see black holes hasn't stopped scientists from identifying them. One of the first celestial objects considered a black hole is an x-ray binary star named Cygnus X-1. Its interactions with its companion star led scientists to believe that it was a very dense object, too dense to be a neutron star or some other type of object. Cygnus has an estimated mass of four to eight times that of the Sun.

An area of future work involves creating equipment to receive the gravitational waves generated in space; this kind of data would provide more direct proof not only of black holes, but also of other astronomical events that are currently unobservable.

Gravitational Lensing

Gravitational lensing occurs when an object has enough mass to bend light. Suppose a distant star emits light to Earth. Now imagine a massive object between this star and Earth. This massive object could be a neutron star, a black hole, or something similar. The high-mass object attracts light from the faraway star and actually bends the straight beams of light. When the bent light makes it to Earth, we see multiple images of the star, depending on its geometry. When astronomers observe multiple images of the same star or galaxy, they know that a high-massed object is somewhere between that object and Earth. If astronomers can't detect the massive object, then odds are good that it's a black hole.

Wormholes

Science fiction suggests that wormholes are connectors between parallel universes, or between two different places in the universe. Astronomical theorists have suggested that wormholes are actually connectors between certain black holes and objects called *white holes*. In other words, if you were to somehow enter a black hole and come out through a white hole, you would have traveled through a wormhole.

QUESTIONS?

If we have black holes, do we also have white holes?
White holes are a purely theoretical construct right now, but the general idea is that white holes might exist at the "other end" of a black hole.

In a rotating or charged black hole it is possible to miss the singularity completely, and instead meet up with a joined white hole and travel through safely. Scientific knowledge in this area is extremely limited, since we have no proof of white holes at all. Even if wormholes do exist, it's unlikely that they would be stable enough to travel through.

Neutron Stars

We know that large and dense stars burn faster than smaller stars. We also know that high-mass stars eventually become black holes and low-mass stars, like our Sun, eventually become white dwarfs. But what do intermediate-mass stars do? As discussed in Chapter 9, at the end of their lifetimes, stars with a mass between 15 and 30 times our Sun's form supernovas with leftover stellar remnants between 1.4 and 3 times the mass of our Sun, then become neutron stars.

If a star's mass is less than 1.4 solar masses, the star eventually stops collapsing due to internal pressure, and the result is a white dwarf. If the central mass is greater than 1.4 solar masses, however, the core keeps collapsing past the point where electrons can support it. Eventually the density and pressure becomes so high that the protons and electrons making up the matter in the star combine to form neutrinos and neutrons. The neutrinos escape, leaving only a superdense object made up mostly of neutrons which supports the star and prevents further collapse. If the mass of the stellar remnant is too large, however (greater than about 3 solar masses), it overcomes the neutron stage and goes on to become a black hole.

Neutron stars are incredibly dense. They contain more mass than our Sun, but are condensed to a typical radius of about 10 kilometers—the size of a typical city! (To put it in perspective, this density would be like packing the mass of all the people on Earth into a volume the size of a sugar cube.) Neutron stars also have huge magnetic fields—the strongest magnetic fields in the known universe. However, neutron stars are so small that they're very difficult to detect, even though they have extremely high temperatures (millions of degrees in their centers). Neutron stars also rotate very quickly, and if they happen to be giving off energy as they do so, we can detect them as pulsars.

Pulsars

Think of pulsars as neutron stars with lighthouse properties. Pulsars give off high-energy radio waves in one direction, but because they are spinning so quickly, you see the flashes only occasionally.

Jocelyn Bell, a graduate student, first discovered pulsars in the 1960s. She was working on a radio astronomy project when she noticed a regular signal in her data. The signal, which turned out to be from outside the solar system, was not a sign of intelligent life elsewhere in the universe, but instead was a radio pulsar.

FACTS

Pulsars give off a steadily pulsing signal that is easily detected on Earth. Scientists first thought these signals could be messages from an intelligent extraterrestrial civilization. Unfortunately, the pulses turned out to be from quickly rotating stars that periodically signal us like beacons.

Pulsars have since been detected with radio telescopes and gamma-ray technology. The Crab and Vela pulsars were two of the first detected through their gamma-ray emissions. To date, over a thousand radio pulsars have been detected through radio transmissions, while only a handful have been found through gamma emissions.

Quasars

Quasar means *quasi-stellar* object, and refers to an entire class of celestial objects that look like stars when photographed, but have high redshifts. When observed with a regular optical telescope, quasars are indistinguishable from stars in our own galaxy—they just appear as unresolved points of light. However, quasars can be distinguished from stars according to their spectra.

Quasars also give off strong radio emissions, and can be detected by radio telescopes. In fact, quasars were first detected by their radio waves. Electrons in the center of a quasar accelerate, and when in a magnetic field, these electrons emit radio waves. The current theory is that quasars are the centers of distant, young galaxies, and may have the highest energy concentration of any object in the universe. Quasars are therefore extremely distant, extremely luminous active radio galaxies.

Active Galaxies

Active galaxies, galaxies that give off various types of energy, include more than quasars. Blazars are another type of active galaxy, and have occasional bright outbursts. Some active galaxies may also have radio lobes or jets, but these structures are usually visible only in radio images.

Active galaxy emissions come from the active galactic nucleus (AGN), which refers to a galaxy with a high-energy core. Quasars are often thought to be active galactic nuclei, extremely energetic centers to distant galaxies. The best way to learn about AGNs is through x-rays, since they can reach the center of many distant galaxies.

SSENTIALS

The twin Keck Telescopes in Hawaii study quasars, as well as other celestial objects in distant galaxies, but many of our best pictures of quasars come from the Hubble Space Telescope.

Seyfert galaxies are yet another type of active galaxy. Seyferts (named for their discoverer, Karl Seyfert) emit low-energy gamma rays that appear to be a continuation of x-ray emissions. This concept might mean that gamma rays come from thermal processes, much like the source of emissions from black holes and other celestial objects. Seyfert galaxies are important for astronomical research because we can learn more about gamma-ray backgrounds in space.

Scientists now believe that all AGNs could result from the same type of physical structure: a large central object (probably a black hole) that is accreting gas from the rest of the galaxy. The various appearances of different AGNs could be due to different viewing angles and different amounts, types, and rates of matter falling onto the central core.

CHAPTER 12

First Steps, First Tools

While your naked eyes are terrific for many types of observing, you'll find it extremely useful to have star maps, binoculars and other simple tools to help you plan your observations and interpret what you are discovering. Planetarium software, magazines, and online resources can provide information, ideas, and insight into your ability to observe.

Star Maps

Johann Bayer (1572–1625) was one of the first astronomers to identify and plot constellations using principal stars, such as Polaris and Altair. Since then, planets, Messier objects, nebulae, and comets have made their way onto most maps. For beginning observers, star maps provide an easy way to find constellations as you become familiar with the night sky.

The earliest star charts were created by a Danish astronomer named Tycho Brahe (1546–1601). He and his team were responsible for the first systematic mapping of the cosmos, an act they performed without sophisticated optical equipment. John Flamsteed (1646–1719), who in 1675 became the first Astronomer Royal with the Royal Observatory in Greenwich, England, produced one of the first star maps created with the aid of telescopes. Flamsteed's ideas were published posthumously in two separate works that detailed the positions of stars and planets: *Historia Coelestis Britannica* and *Atlas Coelestis*. These works formed the definitive basis for future astronomical research, and are a critical part of astronomy's history.

Before actually using your star map, be sure to pick a clear and starry night. Then simply go outside and look up! If you see something that looks like a constellation, turn your star map around to identify it. On star maps, the circular edge represents the horizon, and the center is directly overhead at the zenith. Use a compass (or Polaris, the North Star!) to figure out which direction you are facing.

SSENTIALS Purchase a star map from a local astronomy shop, trading-card store, or science museum. If you plan on observing without the aid of binoculars or a telescope, look for an all-sky map that shows the brightest constellations for your location (latitude) on the day and time you've chosen.

It will probably help to look for the brightest constellations first—ambient and city lights may make the less-bright constellations nearly impossible to see. Remember that constellations, like most things in the cosmos, are considerably bigger than they might appear on your map.

Another approach is to choose a constellation you want to identify. Rotate your star map until the constellation you've chosen looks right side up. You will notice that the other constellations on the map look funky and out of place—that's fine, since you orient the map to the constellation you're focusing on. When starting out, you might want to hold the star map above your head and look up. If you hold it down as you would a conventional book, you might confuse your orientation.

Planispheres

The problem with a star map is that you need a different one for each month and location on Earth. A planisphere is a more generalized star map that can be changed to show the sky at various dates, times, and locations. In astronomy, a planisphere is a projection of the celestial sphere onto a plane. Planispheres contain a moveable, usually circular, overlay that demonstrates which stars and constellations are visible for a particular date and location.

Early History

The sixteenth century cartographer Gerardus Mercator (1512–1594), famous for his projection map of the world, is generally attributed with creating the first modern planisphere. His twelve copper plates engraved with maps of the sky went on to become the reference for later interactive maps. The concept of using a rotating wheel to identify heavenly bodies goes back much further, however. In the first century B.C., Roman engineer Vitruvius wrote about engraved plates with rotating wheels used to mark the rise and set of planets and stars.

The purpose of today's planispheres, or star wheels, is to calculate stars' positions. Because Earth is constantly in motion, the locations of stars and constellations change all the time. A constellation in one part of the sky in August will not be in the same place in November. By using a planisphere, you can rotate the dial to see the correct positions of all the stars throughout the night and throughout the year.

How to Choose a Planisphere

Planispheres are available at most astronomy and telescope shops, and from some science museums. When you're making your selection, consider the following:

- **Legible print:** Avoid the glow-in-the-dark models because the paint is generally thicker and harder to read, and you probably don't want to be squinting at it while you're trying to find constellations.
- **Location:** Less expensive planispheres will probably correspond to only one latitude, so if you require multiple latitudes, make that a priority while you're shopping.
- **Construction:** Planispheres made out of thin card stock with a clear plastic rotating wheel are quite inexpensive and accurate.

Before you begin, turn the wheels on the planisphere to correspond to the date, time, and latitude of your observation. Hold the planisphere in front of you as you face the horizon. Twist it around so the map edge labeled with the direction you're facing (north, south, east, west) is down. The horizon on the map now appears horizontal, and corresponds with your real horizon. Finally, compare the stars on the map to the ones you see in the sky. The circular edge of the planisphere is the horizon; if you were to turn around in a circle, theoretically you would see everything as it is drawn on your planisphere.

Be sure you take distortion into account—everything in the southern sky will seem stretched sideways if you are using a planisphere designed for the Northern Hemisphere, and vice versa. The planisphere is necessarily small, since it has to fit in your hands and be easy to use, but the sky is enormous. Remember to expand your field of view and expect constellations to appear large. An inch or two on your planisphere might correspond to a huge distance across the sky.

Binoculars

Once you are comfortable using a star map or planisphere, it may be difficult to see everything you would like with only your naked eyes. Although basic constellations are relatively easy to find, you will want to locate more stars and fainter constellations. The time has come to make the next step into observing: purchasing binoculars!

Binoculars typically have three parts: a front lens, a rear lens, and a prism in the middle. The front lens (called the *objective lens*) is responsible for gathering incoming light and focusing the image. The rear lens, or the eyepiece, magnifies the image. Since an upside-down image will be produced when we look through the lens, a prism inside the binoculars inverts the image so that as we look through the eyepiece, we see things right-side up.

What to Look For

Given the variety of binoculars on the market, it's easy to let price, magnification, brand name, and the range of options affect your final decision. When purchasing optical equipment, you want to avoid cheap toys. Compromising on quality will also compromise your ability to observe the cosmos. Quality binoculars are usually available at a telescope or camera store, or a high-end sporting-goods store. Try to find a knowledgeable salesperson, and consider his or her recommendations carefully. Remember that you want binoculars for nighttime use, not bird-watching (although the right pair can be used for multiple purposes).

First, consider the optics—the lenses and viewing mechanisms that enable you to see. Size and weight are your next major concerns. You won't enjoy your observing experience if your arms get tired or your binoculars become unwieldy after several minutes. If you are set on getting large binoculars, consider getting a tripod to support and steady them.

FACTS

Each pair of binoculars comes with a set of numbers, such as 8 x 50. The first number refers to the magnification, or power; the second number is lens size, or aperture. In this example, objects will appear eight times closer than what you would see with the naked eye, and the lenses are fifty millimeters in diameter.

Less expensive binoculars may come already focused and cannot be adjusted. Look for binoculars that come with either a center focusing ring, or eye rings that focus independently. Center-focus binoculars typically come with the right eye individually focusable. That's a convenient feature if you happen to be watching moving targets such as wildlife, but they are often more expensive. Since the stars will rarely be moving fast enough for you to need to monitor their movement, a pair with individual focus rings will probably be sufficient.

FIGURE 12-1:
7 x 50
binoculars

(refer to page
279 for more
information)

© 2001, www.artoday.com

Higher-powered binoculars are useful for viewing certain objects, such as planetary moons and star clusters, because they provide substantial enlargement. They do, however, cause a narrower field of view, and often need to be held on a tripod because they are more difficult to hold still. While you can mount most binoculars on a regular camera tripod, or sit in an armchair to steady your elbows, your freedom of movement will be somewhat restricted. If you are planning to simply stargaze with binoculars lars in hand, the usual recommendation is a power of no more than ten.

A high aperture is also preferable, depending on what you are looking for. Bigger lenses let more light through, so stars will appear brighter.

Larger lenses are heavier, however, so before making your decision, try several pairs to see what you can comfortably hold.

Anyone who regularly wears glasses will need to remember to check a new pair of binoculars for eye relief, the distance the viewer's eyes need to be from the binocular eyepiece in order to take in the entire field of view.

Most binoculars have antiglare coatings on the lens' surfaces. More coating means less reflection, so more light will be transmitted through. Coating and price usually go together, so expect a pair of well-coated binoculars to cost more than a pair with few coated surfaces.

How to Choose the Right Pair for You

With this knowledge in hand, how should you go about selecting the right pair of binoculars? The first and most important step you can take

FIGURE 12-2:
Center wheel focusing binoculars

(refer to page 279 for more information)

© 2001, www.arttoday.com

in finding the correct set of binoculars is to go to a store that has a good selection and try out several different pairs. Compare weight, general workmanship, and ease of use. Make sure you can easily work all the focusing devices, and that everything appears to be firmly in place. Shine a light into the barrel and check for dirt on the inner lenses (which usually cannot be cleaned). Repeat this test looking directly into the eyepieces.

If at all possible, test-drive your binoculars at night; many astronomy stores will offer this possibility. Go with the pair that meets your needs and price range, feels good in your hands, and will last as long as possible.

Planetarium Software

Astronomy software presents one of the most recent advances in using technology to enhance your observing experience. Using one of several different software programs, you can locate constellations, learn about stars and planets, or pinpoint where certain planets are at any given moment. Planetarium software can guide your observing, and make the most of your astronomical experiences.

Starry Night

Starry Night, produced by Space.com Canada, developed a PC-based package that allows astronomers to learn about the sky. The software comes in three different versions:

- **Starry Night Beginner** uses simple point-and-click searching to identify planets, stars, and constellations.
- **Starry Night Backyard** allows users to view roughly a million different stars, with convenient identification and learning techniques.
- **Starry Night Pro** demonstrates star fields and over 19 million celestial objects; it also allows users to track comets, satellites, and asteroids.

For more information on Starry Night products, visit their Web site at *www.starrynight.com.*

Voyager

Voyager III, developed by Carina Software, is a Macintosh- and PC-based series. Voyager III uses data from the Hipparcos and Tycho star catalogs to provide precision identification for over a million stars. Users can manipulate star chart data by contrast, magnitude, color, and brightness to create highly customizable viewing data. Explore planets, galaxies, constellations, and everything else the cosmos has to offer. Their Web site, *www.carinasoft.com,* has additional information.

Are there good astronomy sites on the Internet?
Internet resources come and go daily. Use a good search engine, and look for sites that are endorsed by reliable associations, such as NASA, Search for Extraterrestrial Intelligence (SETI), or Students for the Exploration and Development of Space (SEDS). Appendix D lists additional resources.

Celestron

Specialized software packages generally make less of a dent in your wallet. Celestron, the manufacturer of powerful telescopes and binoculars, has developed a particularly good package: The Sky; Level 1—for Celestron. Use this software to generate and print sky maps. The software presents a useful alternative to books if you are looking for multiple maps. Their Web site, *www.celestron.com,* has more information on all of their products.

Cartes du Ciel

Cartes du Ciel is a free software program that allows users to create their own sky charts. The program provides basic locations for planets, comets, asteroids, and other celestial bodies. While the price is probably one of the most appealing attributes of this program, it can be quite useful for creating customized star charts, especially if you don't happen to live near a telescope shop or science museum. The software can be downloaded from *www.stargazing.net/astropc/.*

Astronomy Magazines

A subscription to an astronomical magazine is a great asset for beginners and amateur astronomers alike. In addition to articles on a variety of tropics, such magazines will let you know about upcoming comets and meteor showers, and other celestial events. Most even include a monthly star chart that you can use in your observing sessions! The two major astronomy magazines in the United States are *Sky & Telescope* (*www.skypub.com*) and *Astronomy Magazine* (*www.astronomy.com*).

Color Photograph Reference

1. Planets of the Solar System, without Pluto
   (photo courtesy of NASA/JPL/Caltech)

2. Ultraviolet image of the Sun, with solar prominence, taken by *SOHO* spacecraft
   (photo courtesy of NASA/JPL/Caltech)

3. Color composite image of the Sun at three ultraviolet wavelengths, from *SOHO*
   (photo courtesy of NASA/JPL/Caltech)

4. The Earth rising over the moon, as seen from *Apollo*
   (photo courtesy of © Digital Vision Ltd.)

5. The surface of Mars, from Mars *Pathfinder*
   (photo courtesy of NASA/JPL/Caltech)

6. Close-up view of ice rafts on Europa, in enhanced-color, from *Galileo*
   (photo courtesy of NASA/JPL/Caltech)

7. Enhanced-color view of Europa, from the *Galileo* spacecraft
   (photo courtesy of NASA/JPL/Caltech)

8. The rings of Saturn in false-color, from the *Voyager* spacecraft
   (photo courtesy of © Digital Vision Ltd.)

9. Saturn's northern hemisphere, from *Voyager*
   (photo courtesy of © Digital Vision Ltd.)

10. Volcanic eruption from Io, in false-color, from *Voyager*
    (photo courtesy of © Digital Vision Ltd.)

11. The Milky Way galaxy
    (photo courtesy of © Digital Vision Ltd.)

12. The center of the Milky Way
    (photo courtesy of © Digital Vision Ltd.)

13. Warped galaxy, seen edge-on. ESO 510-G13, from Hubble Space Telescope
    (image courtesy of NASA and The Hubble Heritage Team (Association of Universities for Research in Astronomy [AURA], Space Telescope Science Institute [STScI], NASA))

14. Near-collision of two galaxies, NGC 2207 and IC 2613, from Hubble
    (image courtesy of NASA and The Hubble Heritage Team (Association of Universities for Research in Astronomy [AURA], Space Telescope Science Institute [STScI], NASA))

15. Spiral galaxy NGC 4414, from Hubble
    (image courtesy of NASA and The Hubble Heritage Team (Association of Universities for Research in Astronomy [AURA], Space Telescope Science Institute [STScI], NASA))

16. Halley's comet, as seen in 1986
    (photo courtesy of © Digital Vision Ltd.)

17. The Whirlpool galaxy, M51, from Hubble
    (image courtesy of NASA and The Hubble Heritage Team (Association of Universities for Research in Astronomy [AURA], Space Telescope Science Institute [STScI], NASA))

18. Globular Cluster M80, from Hubble
    (image courtesy of NASA and The Hubble Heritage Team (Association of Universities for Research in Astronomy [AURA], Space Telescope Science Institute [STScI], NASA))

19. Gaseous pillars in the Eagle Nebula, from Hubble
    (photo courtesy of © Digital Vision Ltd.)

20. A stellar explosion—shockwaves from a supernova, from Hubble
    (photo courtesy of © Digital Vision Ltd.)

21. A Nebula, from Hubble
    (photo courtesy of © Digital Vision Ltd.)

22. The Orion Nebula, from Hubble
    (photo courtesy of © Digital Vision Ltd.)

23. The Keyhole Nebula in Carina, from Hubble
    (image courtesy of NASA and The Hubble Heritage Team (Association of Universities for Research in Astronomy [AURA], Space Telescope Science Institute [STScI], NASA))

24. Planetary Nebula NGC 6751, from Hubble
    (image courtesy of NASA and The Hubble Heritage Team (Association of Universities for Research in Astronomy [AURA], Space Telescope Science Institute [STScI], NASA))

25. Planetary Nebula Mz 3, the "Ant Nebula," from Hubble
    (image courtesy of NASA and The Hubble Heritage Team (Association of Universities for Research in Astronomy [AURA], Space Telescope Science Institute [STScI], NASA))

26. Death of a star
    (photo courtesy of © Digital Vision Ltd.)

27. The Tarantula Nebula, from Hubble
    (image courtesy of NASA and The Hubble Heritage Team (Association of Universities for Research in Astronomy [AURA], Space Telescope Science Institute [STScI], NASA))

28. Iras II
    (photo courtesy of © Digital Vision Ltd.)

29. Einstein Cross, example of gravitational lensing
    (photo courtesy of © Digital Vision Ltd.)

30. Supernova 1987A
    (photo courtesy of © Digital Vision Ltd.)

31. Black Hole
    (photo courtesy of © Digital Vision Ltd.)

5

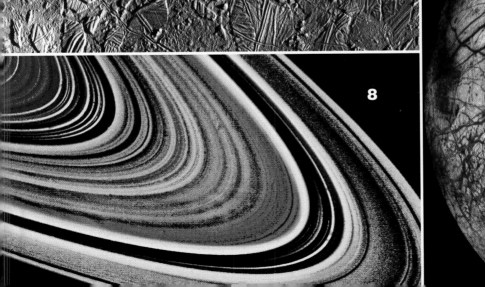

6

8

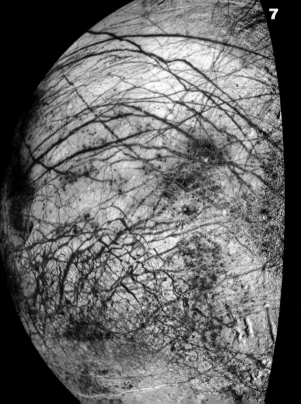

7

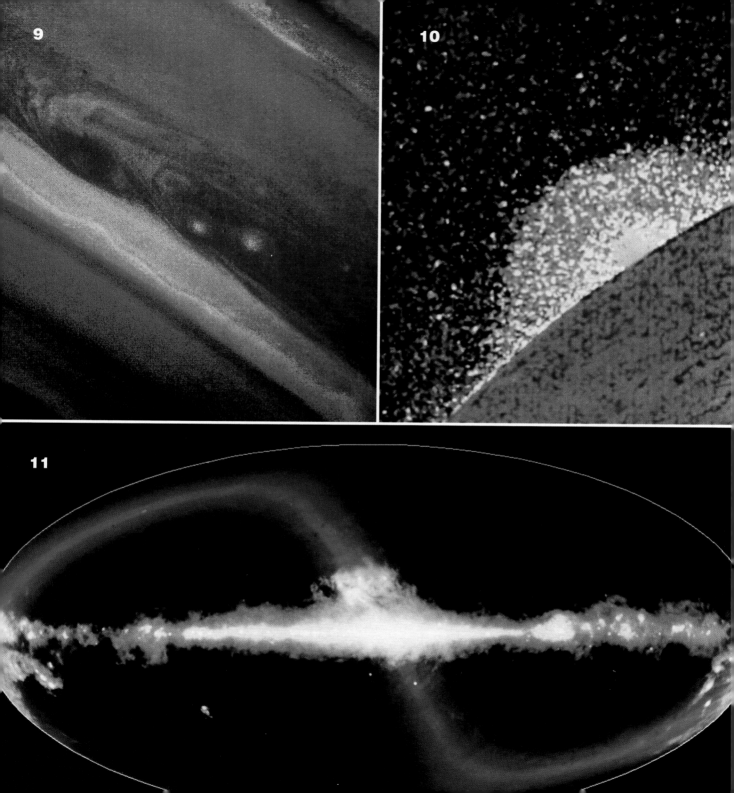

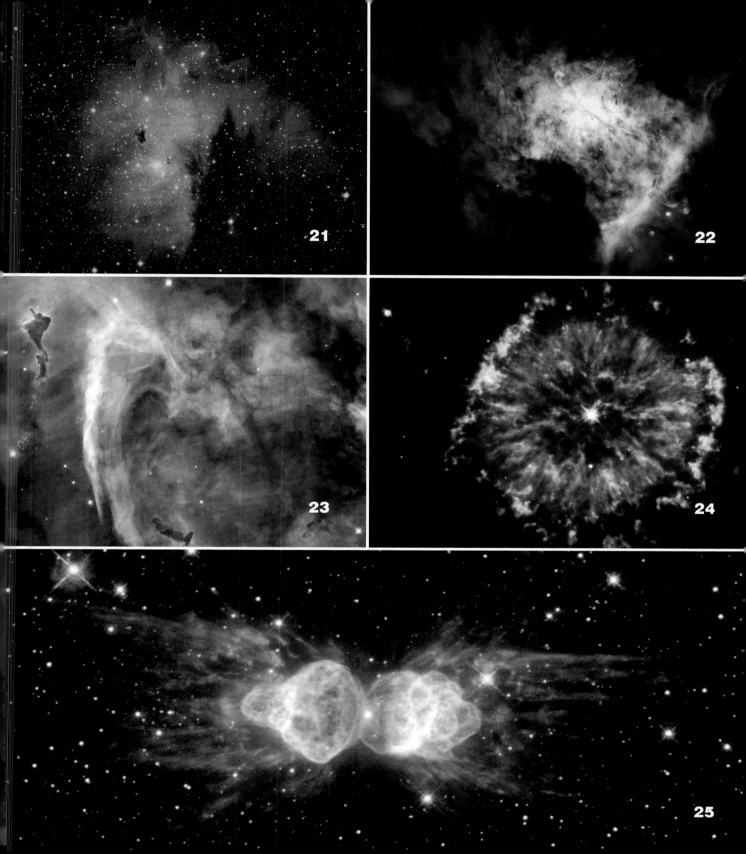

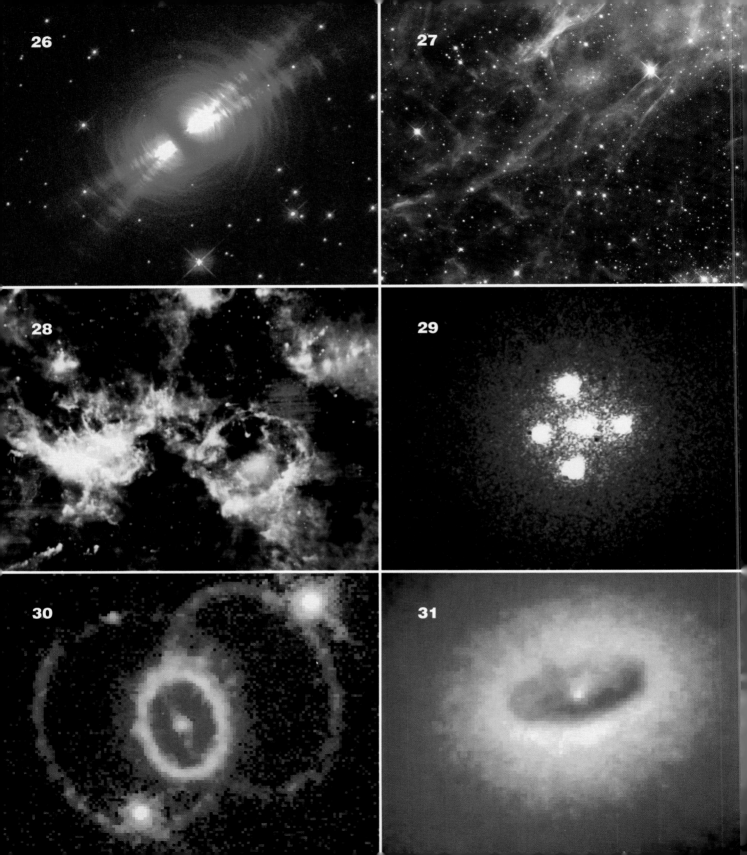

CHAPTER 13

Telescopes: Advanced Tools for Advanced Viewing

In 1609, Galileo peered through a telescope and identified heavenly bodies. Today, we realize that the telescope led to discoveries and scientific achievements far beyond what he could have imagined. Telescopes had certainly been in use before Galileo, but his application of optics to the heavens inspired new research, new discoveries, and the eventual availability of telescopes to individual consumers.

History

Telescopes as we know them today were originally built for use by eyeglass makers. A Dutch optician named Hans Lippershey (1570–1619) created one of the first usable telescopes in 1608. Legend has it that children playing with lenses in his eyeglass shop noticed their property of magnification when used together, and thus was born the first telescope. Galileo followed suit the next year with telescopes that were suitable for skyward gazing, creating eight-power and twenty-power instruments.

Choosing a Telescope

Telescopes are advanced tools, but like binoculars, they come in several sizes and types, and have different purposes. They also come in a wide range of prices, depending on the quality and technology involved. Telescopes will be most useful to you after you've learned basic constellations, and the positions of solar-system and deep-sky objects. Be sure you've spent plenty of time practicing with binoculars and star maps before investing in a telescope.

What to Look For

Many stores sell telescopes, but you will be best served if you look in an astronomy or science shop. Department stores and toy stores often sell telescopes, but as you might expect, these are often toys and are not suitable for serious observing.

Many factors come into play when selecting your first telescope. In order to prioritize your needs, consider the following:

- Do you live in the city, with a lot of ambient light; or out in the country, where the night sky is dark?
- Do you have a large back yard, where you could mount a telescope permanently, or do you live on the top floor of an apartment building, and need something lightweight and portable?
- Are you prepared to carry a significant amount of weight up and down stairs?

- Can you transport large tubes?
- Do you want to study planets, stars, constellations, or galaxies?
- Do you plan to simply observe the cosmos, or do you want to track specific objects across the sky?

After considering the questions above, you can offer your priorities and criteria to whomever is helping you make your decision. As with any purchase, comparison shopping is critical to finding the best telescope for your needs.

A decent telescope will probably cost at least $200 to $300. The store you choose should have a good selection and professional service to ensure you make the best decision. There are a number of reputable mail-order and online telescope dealers, but purchasing a telescope sight unseen is probably not a good idea for your first instrument. When you go shopping, try to take a friend; having a second pair of eyes will help you make the most discriminating choice.

Choosing a Good Fit

Again, weight and ease of setup are primary factors. More powerful telescopes tend to be heavier, so if you need help carrying and setting up your telescope, you might use it less often. Talk to professionals and practice setting up several different models to see which fits your abilities and lifestyle the best. Make sure the telescope you select comes with a case, if practical (depending on the size of the telescope)—it will make carrying easier, and will help protect the optics during transportation and storage. If a case is not included, purchase one separately.

Telescope Brands

In terms of telescope brands, there are two major players. Both companies make extremely high-quality devices, as do a number of smaller telescope manufacturers. Celestron is one of the most prominent manufacturers in the industry because its materials and design are high quality. Celestron manufactures telescopes, binoculars, spotting scopes, microscopes, and other optical equipment.

Check the warranty period on the telescope you select. Although it may seem trivial at first, you'll appreciate the protection should you encounter a mechanical or other failure during the first few months.

Meade also manufactures high-quality telescopes. Its full product line includes binoculars, microscopes, and CCD systems. (A CCD, or charged coupling device, is a computer chip with a grid of embedded light-sensitive detectors.) Meade started as a very small company in 1972, and prides itself on providing quality products and excellent customer service.

Types of Telescopes

Once you know what *you're* looking for with regard to size and capability, you'll need to become familiar with what your options are. There are three basic classes of telescopes: reflecting, refracting, and catadioptric.

Reflecting

The reflecting telescope, first designed by Isaac Newton in 1672, uses a concave mirror to gather light and is the most common telescope for amateur astronomers. The optics are high-quality, yet the construction is fairly simple. The open design generally keeps moisture from forming on the interior surfaces, which is a tremendous benefit on overnight observing trips. The open design can be a downfall, however, because dust can get inside the telescope and you need to take special care when cleaning it.

Refracting

The refracting telescope, like Galileo's, gathers light through a lens, then delivers the light to the eyepiece. The lens focus is usually near the front of the telescope. Refractors typically have thin, long tubes that let you focus on most objects. Prices are generally more affordable for this type of telescope, but can range from cheap to extremely expensive. Refracting telescopes are usually well constructed. A minimum of moving parts and a sealed interior reduce the likelihood of breakage and expensive servicing. On the other hand, refracting telescopes tend to have small apertures (5 inches or less), which makes it hard to view galaxies and other deep-space objects.

Catadioptric

The catadioptric telescope, or compound telescope, uses both mirrors and lenses. Compound telescopes are most useful for astronomers interested in using electronic tracking for stars and planets, and many catadioptric telescopes are computerized. The short tubes are convenient for packing and travel, and mount readily to platforms for photography and computerized tracking. The most popular models are the Schmidt-Cassegrain and the Maksutov-Cassegrain.

Anatomy of a Telescope

Telescopes, like binoculars, are composed of many different parts. One of the most significant elements is the aperture, or the diameter of the lens or mirror. Larger apertures let in more light from distant objects, so you will be able to see more objects, more clearly. Very large apertures also increase the telescope's weight, so you must find a balance. For deep-sky viewing, look for an aperture of at least 4 inches. Six inches or greater will make observing more pleasurable and productive—ensuring that your new telescope will get plenty of use!

FIGURE 13-1:
A typical reflecting telescope

(refer to page 279 for more information)

© 2001, *www.arttoday.com*

Power

Power, or magnification, is another factor that comes into play. In the case of telescopes, *focal length* refers to the distance that light needs to travel inside your telescope. Focal length contributes to the overall power of a telescope; magnification is the ratio of the telescope's focal length to that of the eyepiece. The power of any given telescope can be estimated by multiplying the aperture size by fifty. A telescope with an aperture of 6 inches, then, has a power of approximately 300.

Eyepieces and Focusing

All telescopes will have some form of eyepiece and focusing device. If you buy one with interchangeable eyepieces, you will be able to observe different planets and stars at different magnifications, getting more use out of a single telescope. In this respect, the power of the telescope you're considering is less important than the eyepieces that come with it, since you can always change your magnification that way.

Beware of department-store telescopes advertised as "125x power"—advertising by magnification alone is a good sign that the instrument you're looking at is more of a toy than a serious astronomical device.

Don't be fooled by advertisements claiming extremely high magnification with a small aperture—larger eyepieces go with more expensive telescopes. An eyepiece diameter of .965 inches is acceptable for most uses, but 1.25 inches will give you greater clarity and granularity. Try out different models, and see the difference for yourself.

Finder Scopes

Good telescopes will have an attachment mounted on the top called a *finder scope*, a smaller scope with a wide field of view that allows you to focus your view on certain stars or planets. Use the finder scope to find objects initially, or to realign your telescope when your subject has gone out of view. The better telescopes have removable, or separate, finder scopes.

Tripods and Mounts

FIGURE 13-2:

Setting up a telescope mount

(refer to page 279 for more information)

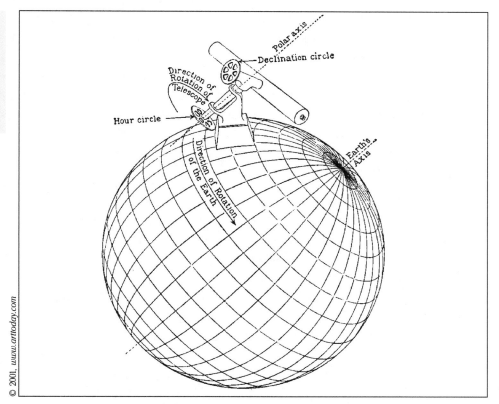

© 2001, www.arttoday.com

You will also need to consider the type of mount or tripod for your particular telescope. There are at least three different types of telescope mounts—altitude-azimuth (commonly called *alt-az*), Dobsonian, and German equatorial.

Altitude-Azimuth Mounts

Alt-az mounts, the most common, involve axes that move both horizontally and vertically. Such tripods will fit a wide range of telescopes, but they require constant attention and relative precision.

German Equatorial Mounts

German equatorial mounts are designed to follow the movements of planets or stars across the sky. They provide a finer range of motion across only one axis, as long as they are set up to be aligned with the north celestial pole. Equatorial mounts are most useful when your goal is to keep a particular body in view for a length of time, or when you plan to use calculations to track an object. If you plan on tracking, be sure to ask about related features for the models you consider.

Dobsonian Mounts

Dobsonian mounts are simplified altitude-azimuth models. Dobsonian mounts are easy to construct and maintain. These mounts use a friction-based design; the telescope maintains position by the friction created between the telescope bearings and the mount. They do not track stars or other bodies, but they are generally the lowest-priced type of mount you can get, and they can support very large tubes. Horizontal and vertical axes both rotate, so if you want to track the progress of any moving body, constant readjustment will be necessary.

When testing a telescope, make sure the object centered in the crosshairs of the finder scope is actually also centered in the main tube—if it's not, the telescope has an alignment problem.

Whichever mount you end up with, make sure it is stable and well-constructed. A good test is to tap the telescope gently; if it does not regain its balance within one to two seconds, you need a sturdier mount. When you are observing at night in a field, especially with other people, you want to make sure your setup is as firm as possible.

Computer-Controlled Telescopes

A new development in telescopes for beginners (and experts!) is the computerized telescope. These telescopes either connect to a computer or laptop, or have a small computer built in. Instead of painstakingly locating an object in the night sky by hand, with a computer-controlled telescope you simply type in the coordinates you wish to go to, or even select the object you wish to observe from a list. The telescope will point itself and the object automatically! Such telescopes are very convenient and appearling for beginners. A few Warnings, however: Computer controlled telescopes can be difficult to set up and align properly. They can also be expensive, and some astronomers feel that they take some of the fun out of learning your way around the night sky. But when used correctly, such telescopes can make observing a lot of fun!

Telescopes at Work: Hubble

Construction on the Hubble Space Telescope (HST) was completed in 1985, and the space shuttle *Discovery* transported Hubble into Earth's orbit in April 1990. Originally scheduled to be launched earlier, the HST was delayed due to factors surrounding the space shuttle *Challenger* disaster of 1986.

Edwin Hubble (1889–1953), former director of the Mt. Wilson Observatory in Pasadena, California, was responsible for the discovery of several distant galaxies. One of his most significant contributions to astronomy was the notion that galaxies farther away from Earth appeared to be moving faster than ones closer to us; his work helped establish the foundation for the Big Bang theory of the origin of the universe.

QUESTIONS?

What is astrophotography?
In the simplest sense, astrophotography is the process of taking a picture of the sky. Some say that connecting a camera and a telescope can be a frustrating endeavor, but the rewards of photographing the cosmos are intense. If you think you might want to get involved in astrophotography, keep that in mind while considering telescopes.

The Hubble Space Telescope is a reflector built from a Ritchey-Chretien Cassegrain model, in which two mirrors combine to focus images over a large field. Hubble uses an optical telescope assembly that consists of two mirrors (measuring 7.9 feet across), supports, and apertures. Hubble combines optics, instruments, and spacecraft systems in one single, extraordinary telescope, including two spectrographs and two cameras. Hubble has successfully monitored planets, stars, and galaxies, as well as other cosmic events.

When Hubble first began sending images back to Earth, problems surfaced—the images were out of focus because the telescope's mirror was too flat along one edge. The solution? Fix it! The space shuttle Endeavor met up with Hubble in December 1993 and modified it by adding a special camera. The Hubble Telescope has taken amazing pictures, including many of the astronomical pictures in this book!

Building Your Own Telescope

So how can you go about building your own telescope? First, find a book or specialized plans on just this subject; there are several different "build your own telescope" books available. Dobsonian models work well for home builders, so you might look for the Dobson type if this is your first project.

Be sure you have the available workspace and the correct power tools before attempting to build your own telescope. Basic materials will include a tube, a box, hardware for a rotating platform, and a mirror. The good news is that most of the equipment is available at local home and hardware stores. The tube can be something as simple as a cardboard concrete-pouring tube, painted black. The box needs to be perfectly square, a more difficult endeavor than nearly any part of the telescope construction. The mirror will probably be one of the most expensive components, and if you plan to grind it into shape yourself, be prepared for weeks of difficult work. With the proper equipment, however, the whole project can be an incredibly rewarding experience.

CHAPTER 14

Visit Space Places

Sometimes there is nothing better than lying under the stars alone, contemplating your place in the universe. Sometimes, though, you might lie there stumped over the difference between the Big Dipper and the Little Dipper. Astronomy is often a hobby and a science enjoyed best in groups. Use all the resources that are available to you, and you'll probably enjoy astronomy even more than you do now!

Clubs

Whether you join an organized club, or have casual star parties, you'll find that people who share your interests don't just foster your enthusiasm in the topic—they raise questions, share knowledge, and keep you current on the newest information.

What Clubs Have to Offer

Astronomy clubs are prevalent in most of the United States, as well as in many other countries. Most clubs have monthly meetings, which are a great opportunity to meet friends in your area for observing runs. Some club Web pages will have links to local weather services—ensuring that you don't end up searching for Cassiopeia in a thunderstorm! With information on meteor showers, comets, and other stellar events, astronomy clubs can help you pick a date and time for optimal observing.

SSENTIALS

Although most astronomy clubs are not free, membership fees are usually modest. They will often include a monthly magazine or newsletter, discounts at various telescope shops and bookstores, and of course access to all their resources.

Most clubs are run by amateur astronomers—people who are interested in astronomy and might have some degree of formal training, but typically have other careers and other jobs. Since local astronomy clubs often have resident experts, they are typically a great source of information on the best local places to observe—whether they have observatories or just dark-sky locations. These tips are especially useful if you are new in town, or just new to astronomy. Clubs also have the ever-useful bulletin board, a place for posting used equipment for sale or trade—a great resource for bargains and advice.

Star Parties

Some clubs hold occasional star parties. These events are one of the best ways to get to know the astronomical community in your area, as

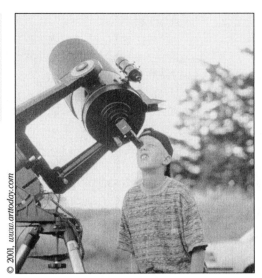

FIGURE 14-1:
Go to
a star party

(refer to page
279 for more
information)

© 2001, www.arttoday.com

well as make terrific observations. At a star party, people bring their telescopes to one location and look at stars together. Clubs often have a dark-sky place that they use, and may have different sites for different times of year. At star parties you can try out different telescopes and learn how to find obscure objects. If you don't own a telescope yet, these events are a great way to try several different kinds, and decide for yourself which one best suits your needs.

Star Party Etiquette

- Use a red-light flashlight, but only when necessary. If you don't have a red light bulb or filter, tape a piece of red cellophane over a regular flashlight. The filter will allow enough light for you to see without disturbing other observers.
- Park your car so that the headlights won't disrupt other observers if you leave early, and always drive slowly. Use just your parking lights when approaching the observing site.
- Let your eyes adapt to the dark before attempting to move around the site, so you don't walk into someone's delicate equipment or trip over power cords that could be tethered to car batteries.
- Find out ahead of time if pets are welcome, and if they are, plan to bring a leash. If you bring a radio, use headphones.
- Always ask permission before using someone else's equipment, and avoid touching the optical surfaces. Treat other people's telescopes respectfully; be careful and make sure your hands are clean. Even something as simple as eye makeup can leave unwanted traces on an eyepiece.

Eventually, you'll learn what's appropriate for your group or club, but certain "rules" are universal. Lanterns, campfires, and other light sources are generally inappropriate because they make observing more difficult for everyone. If you need to turn on a regular light for some reason, tell everyone first to make sure you don't ruin a good photograph. Be as considerate as possible, and you will get the most out of your experience.

Planetariums

Planetariums and observatories provide amazing opportunities for learning about the solar system, and viewing celestial events. A planetarium is usually a room in a building or museum that has a large projector; images of the heavens, such as constellations and other objects, are projected onto the walls and ceilings. Such rooms are often domed for greater effect. The audience sits in reclining chairs and looks at the ceiling as the show begins. A special type of projector is used to simulate the appearance and motion of the night sky on the dome above you. These projections are often combined with slides, music, and other visual effects to create a great show.

FIGURE 14-2:
The Flandrau Planetarium at the University of Arizona in Tucson

(refer to page 279 for more information)

© 2001, www.arttoday.com

FIGURE 14-3:
An early version of the Zeiss star projector

(refer to page 279 for more information)

© 2001, www.arttoday.com

During a planetarium show, a live guide or recorded tour will explain what you are looking at, and will give viewing tips and hints. The Hayden Planetarium at the American Museum of Natural History in New York is one of the finest examples. The top half consists of a space theater using a Zeiss star projector, while the bottom half offers a recreation show of the Big Bang. Many science museums have planetariums, and you can use the list included in Appendix D to find one in your area.

Observatories and Science Museums

FIGURE 14-4:
Kitt Peak National Observatory, near Tucson

(refer to page 280 for more information)

© 2001, www.arttoday.com

Unlike planetariums, observatories are rooms or buildings equipped for actual astronomical observation. They usually have telescopes and computer-controlled viewing and imaging equipment. Most have wide fields of view, maximizing the amount of visible sky. Professional observatories are usually located on mountaintops in dry, clear climates. The Kitt Peak National Observatory in Arizona is one of the most famous. Located near Tucson, Arizona, this observatory has many different telescopes and imaging systems, and you can arrange a guided tour of the telescopes during the day if you plan ahead. The Kitt Peak visitor's center has a smaller telescope for amateur astronomers to use!

FIGURE 14-5:
The Smithsonian National Air and Space Museum in Washington, D.C.

(refer to page 280 for more information)

© 2001, www.arttoday.com

FACTS

The National Air and Space Museum in Washington, D.C., has spacecraft exhibits and models from both manned and unmanned spacecraft. You can sit in an Apollo capsule, or you can view and touch Moon rocks and other artifacts from space.

Science museums have a broader focus and a wider range of exhibits than either a planetarium or an observatory. Most museums have at least a few exhibits on space and astronomy. Some have IMAX movies that show space-related movies on a very large screen; others have telescopes on the roof where you might be able to attend a star party. If your town has no planetariums or observatories, a science museum would be a good choice for learning more about astronomy.

CHAPTER 15

Make Your Own Scientific Discoveries!

One of the most exciting things a budding astronomer can do is be the first human being on the planet to see a comet or asteroid. To make such a discovery takes a lot of hard work and persistence, but the payoff is tremendous. In fact, most comets and many asteroids are discovered by amateur astronomers even today. Amateur astronomers also make important observations of meteor showers and other transient phenomena.

Discovering a Comet or Asteroid

Comets are the only bodies in the solar system named directly after their discoverer, and most are discovered by people called comet hunters. Comet hunters are dedicated, generally amateur, astronomers who spend many hours each night searching the sky for any changes from their previous observations. Amateurs also usually find asteroids, though asteroids are harder to find because they aren't as bright as comets. An asteroid discoverer gets to select a name for the newly discovered object, too, but unfortunately it can't be his or her own name.

FIGURE 15-1:
An asteroid
trail in
Centarus

(refer to page
280 for more
information)

Courtesy of NASA and The Hubble Heritage Team (STScI/AURA)

Tools and Equipment

The first thing you need is regular access to a telescope. Make consistent observations of the same parts of the sky every time you observe. The more familiar you are with what an area usually looks like, the better your chances of noticing something different. Also, familiarize yourself with known stars, planets, and other objects in the areas you are observing.

Next, obtain a definitive catalog of existing celestial bodies. *Sky Atlas 2000.0,* by Wil Tirion and Roger Sinnott (Cambridge University Press, 1998), is the standard catalog of the heavens with twenty-six charts, at a resolution of 8.2 millimeters per degree. Approximately 2,700 objects are represented and labeled, and the charts are easy to see in both daylight and under red light.

Procedure

Once you sight something that you believe might be a new comet or asteroid, the first thing you will need to do is look it up in your reference atlases and make sure it's not an existing star or other object. Check the deep-sky catalogs, and verify that there is nothing in the region you are studying.

The Appearance

Once you've attempted to verify that what you're looking at is new, you should make a field sketch. This is a sketch made while you are looking through your telescope. While you can take a quick look through an eyepiece for regular viewing, sketching requires a more focused and concentrated observing session.

ESSENTIALS

If you believe you have discovered a new comet, don't get too excited until the astronomical community has had time to verify that it really is new. Many comets have actually been discovered previously, then lost.

Astronomical sketching really requires only a piece of paper, a red light, and a hard drawing surface (such as a clipboard)—drawing on soft grass will seriously diminish the quality of your record. You can also construct a light box to help you see your drawing in the dark. To make a simple light box, install a couple of red LEDs (light-emitting diodes) under a translucent surface, then encase the setup in a thin box. This arrangement will let you trace over an existing star map in the dark, and might help pinpoint your discovery when you get back home.

What is the point of making field sketches?
Field sketches give you something to look back on as a reminder of your observations. If you don't have a camera attached to your telescope, they are an excellent way to keep a record of what you have seen.

If you are familiar with the area of the sky where you're going to be observing, bring a photocopy of a star map to start the drawing. Alternatively, begin the sketch ahead of time and simply record your findings as you see them. Either way, creating a field sketch will be essential to tracking your discovery and presenting it to the astronomical authorities for verification and identification.

The Stats

In addition to the sketch, make brightness measurements of your findings. One of the easiest ways to estimate the brightness of a possible new comet or asteroid is to check an atlas for the brightness of a known star, and compare it to your new object. The method is clearly lacking in accuracy, but can serve well for initial estimates. (Video meters can also take actual measurements of celestial body brightness, but amateurs aren't likely to have such tools.)

Reporting Your Discovery

Once you are fairly certain that what you have is a new comet or asteroid, get in touch with the Central Bureau for Astronomical Telegrams. CBAT is located in the Harvard-Smithsonian Center for Astrophysics at Harvard University, in Cambridge, Massachusetts. The bureau has an enormous database of all known celestial objects. When it receives a new observation, CBAT checks it against where everything in the database would be at your recorded date and time, to make sure your discovery hasn't already been found by someone else.

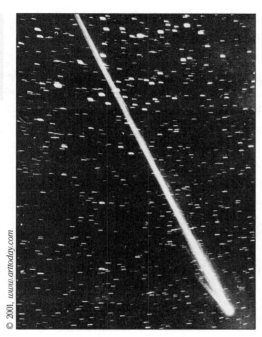

FIGURE 15-2:
A comet's tail

(refer to page 280 for more information)

© 2001, www.artoday.com

If your discovery is indeed something new, CBAT sends an alert to observers around the world for someone to confirm your discovery. With a series of confirming observations, CBAT computes an orbit for the object, mainly to see if, and how close, it may come to Earth. The more observations CBAT receives, the more precise the orbit data will be.

Most comets are discovered when they are fairly far from the Sun, so typically all you will see is a bright spot, or smudge, that should not be there. You won't see a comet's tail on first observation. The tail does not develop until the comet approaches the Sun, which heats up the ice of the comet and vaporizes a portion of it.

Comet Hale-Bopp

Comet Hale-Bopp is an excellent story of two people discovering a single comet. Alan Hale was a serious amateur astronomer who logged many hours searching the skies. July 22, 1995, was his evening to remember. In New Mexico at the time, Hale was observing two previously found comets when he observed a brightness in Sagittarius that he knew had not been there before. He watched the bright object, verified that it was both new and moving, and reported it to CBAT.

FACTS

Comets, asteroids, novas, and similar bodies can currently be reported by sending an e-mail to *cbat@cfa.harvard.edu.* This address could change in time, so be sure to check the Internet for the latest information.

Thomas Bopp, also an astronomer, did not own a telescope when he made his observation. Fond of star party gatherings, he made many observations at dark-sky parties with friends. The night he first saw comet Hale-Bopp, he was observing with friends in Arizona. Using a Dobsonian reflector telescope, Bopp also noticed a glow in Sagittarius that did not appear to be on any charts. He made some initial calculations, sent a telegram off to CBAT, and soon learned that he had codiscovered a brand new comet!

Auroras

FIGURE 15-3:
An artistic rendering of an aurora

(refer to page 280 for more information)

© 2001, www.arttoday.com

Auroras are another astronomical event that you can observe from Earth. You don't discover auroras, per se, but they are an amazing sight to see. Storms on the Sun send off solar flares of radiation and high-energy particles. When those particles hit Earth's magnetic field, the glow that results is an aurora. Auroras are primarily a polar phenomenon, and are most common in places such as Antarctica, Scandinavia, Russia, Alaska, and Canada.

FACTS

Auroras are commonly called *northern lights*. Aurora borealis is a northern aurora, while aurora australis occurs in the south. Auroras grow stronger during solar storms, or periods of high solar activity.

The colors, bright reds, greens, and yellows, depend on the type of atom being struck by the high-energy particles. Auroras generally occur between 50 and 100 kilometers above Earth, and have been documented for at least 3,000 years. Because the particle collisions produce x-ray light, which is not produced by Earth's atmosphere, auroras are studied through x-ray telescope observations. If you are ever traveling in a country close to the north or south polar regions, make sure to look up at night!

Recording a Meteor Shower

Discovering a new comet is an incredible way to get your name into the history books. However, there are plenty of other ways you can enjoy astronomy and make valuable contributions to science. Recording a meteor shower can be of immense help to astronomers, and it's fun! Although videos and photographs of meteors usually yield more accurate data, the sheer quantity of naked-eye observations from amateur astronomers lends them credibility in the scientific field.

Selecting a Shower: Leonids and Perseids

For starters, pick a good meteor shower to observe. Meteors, the result of dust particle collisions with Earth's atmosphere, travel so quickly that they heat up, causing them to glow; this phenomenon is what we often call *falling stars*.

Choose a major shower, such as the Leonids or the Perseids. The Perseids occur every year around August, and are a spectacular show. The best times to see the Perseids is during a new Moon, eliminating the Moon's light from the night sky. Check astronomy magazines or Web sites for information about when to best view the showers.

The Leonids, so named because they appear to be coming from the constellation Leo, were first observed in 1833 and peak every thirty-three years. The parent comet, from which the Leonid meteor shower originates, is called Tempel-Tuttle. The last major storm occurred in 1998 to 1999, so be on the lookout for the next one in about thirty years. In the meantime, you can see the Leonids every November in smaller minor showers.

Preparing for Your Observation

When you're actually observing, how can you be sure that what you're seeing is part of a meteor? For starters, do your homework and determine where your intended meteor is originating—if you plan to observe the Leonids, make sure you can identify Leo. Meteors are generally one of two types: sporadic (those that originate from a random place) and radiant (those coming from a special area, such as the Perseids or Leonids).

Here are a few considerations in planning your observation:

Optimize your viewing. Pick a shower that will have at least twenty visible meteors per hour. Choose a dark-sky site, and try for a clear night (minimal haze, smog, or clouds). Make it fun—go on a camping trip, and take some friends!

Dress appropriately. If you plan to be out at 3 A.M., wear enough warm clothes to last through the shower's peak. Warm fluids and gloves will also help keep you happy and warm during winter observations.

Be prepared. You will probably be observing for a few hours, so consider bringing a lawn chair or something comfortable for lying on the ground. Blankets are essential, as are red-light flashlights, paper with clipboards, and pens or pencils. If you live in an area with mosquitoes or other summer pests, bring repellant so you can observe undisturbed.

Keep a record. Remember to bring a watch for timing your observations; sometimes a shortwave radio tuned to a time signal can be useful as well. Some astronomers bring a tape recorder and orally describe each meteor they see, but tapes can fail so have a backup plan. Check your equipment ahead of time, bring extra batteries, and be sure to use a new tape. If you plan to use a telescope or camera, bring the appropriate equipment.

Perhaps most important, make sure you're well rested! Don't go on an observing run if you were up all night the night before. An afternoon nap can go a long way toward making you more alert as the wee morning hours approach.

Star maps and compasses are helpful if you're new to this, but perhaps the best tool on your first meteor-observing trip is an experienced friend. Not only will you have someone to help identify the constellations with you, but you'll also enjoy the company.

Documenting Your Observations

Mark down the time you start observing and the time you finish. Also take note of cloud cover or other atmospheric conditions. One of the most popular ways to record meteors is called the counting method. As soon as you see a meteor, write down the time, estimate the brightness, note the color (if any), and see if there is an observable tail. Use a tape recorder if it's easier for you. You should definitely record all meteors you see, making notes as to their relative sizes and brightness.

As the night gets darker and your eyes become adjusted to the dark, you will be able to see more. When you record the brightness of a meteor, you compare it to your limiting magnitude (the brightness of the faintest star you can see without a telescope), and this method gives a relative frame of reference.

Reporting Your Data

There are several groups to which you can report your data. One of the primary organizations is the American Meteor Society (AMS), a scientific group studying meteor astronomy. The AMS has been gathering professional and amateur data since 1898. If you plan on submitting your observations, the AMS requires the following information: your location and how cloudy it was the night you observed, the direction you were facing, and your altitude. Most people report meteor data to the AMS either every few months or once a year.

FACTS

The American Meteor Society has a variety of information, including sample observing forms (and instructions), on their Web site, ✍ *www.amsmeteors.org.*

The International Meteor Organization also coordinates the findings of amateur astronomers around the world. Its fundamental belief is that as people from different countries and continents supply meteor information, our knowledge as a world becomes more complete.

Photographic and Video Observation

It's easy to observe meteor showers with the naked eye, but photographic observation is very exciting and can yield more accurate data. From a series of photographs, you can measure height, size, and even the speed at which the meteors are traveling. The most important feature in a camera to be used for photographing meteors is the ability to take long time exposures. Your camera must be able to handle both humidity and frost without fogging up; sometimes a heater will be required to keep dew from forming. Use a high-sensitivity film, such as ISO 3200/36.

Video observation is an alternative to photographic, but is generally more complicated and more expensive. Features such as meteor light curves can be detected because video cameras have dramatically improved limiting magnitudes, and can "see" more meteors. Unfortunately, the initial investment for video observation of meteor showers usually knocks this option out of the amateur loop.

Observing Variable Stars

Observing variable stars is another way in which amateur astronomers can make significant contributions. Variable stars change brightness periodically. More than 30,000 variable stars have been discovered and cataloged, making their observation a popular occupation. In observing variable stars, we learn about their behavior and periods of variation, in addition to increasing our general knowledge of stars.

There are several different types of variable stars:

Pulsating Variables: The actual layers of the star expand and contract, giving off varying amounts of light. The Cepheids have a period of between one and seventy days; pulsating variables with longer periods can go up to a thousand days.

Cataclysmic Variables: Also called *eruptive variables*, these stars' brightness is altered suddenly and violently by occasional explosions from within the star. Supernovas are cataclysmic variable stars. Novas typically have periods of between 1 and 300 days.

Eclipsing Binaries: These are binary stars whose orbital planes lie close enough to our line of sight that one sometimes passes in front of the other, making it look less bright to an observer on Earth. The period of this type of star can be several minutes or several years.

Amateurs can provide important data to the observation of variable stars because more eyes are better than fewer, and people in different parts of the world can often make crucial observations that a professional might miss. Some star periods vary over a period of hours, making them fairly easy to observe, but others vary over periods of weeks or months—in these cases, amateurs are needed to provide more recorded observations.

Observing Transits, Occultations, and Eclipses

Another category of interesting astronomical events includes transits, occultations, and eclipses. These phenomena give us important information about the orbit and brightness of both celestial bodies involved. Careful timing of such events is critical because it leads to precise information about planetary orbits.

QUESTIONS?

How are an eclipse, an occultation, and a transit different?
An eclipse occurs when one body passes into the shadow of another; an occultation is the covering of a small celestial body by a larger one; and a transit is the passing of a small celestial body over a larger one. For example, when studying the satellites of Jupiter, an eclipse occurs when a satellite passes into Jupiter's shadow; an occultation occurs when the satellite passes completely behind Jupiter; and a transit takes place when the satellite passes in front of the planet, between Jupiter and the observer.

FIGURE 15-4:
A solar
eclipse

(refer to page
280 for more
information)

Courtesy of NASA/Johnson Space Center [JSC]

Eclipses, the best known of these events, are special cases of a transit or occultation. A solar eclipse occurs when the Moon moves in front of the Sun, as seen from Earth. A total solar eclipse, during which the Sun is completely covered by the Moon for a few short minutes, is a rare and spectacular celestial event. If a total solar eclipse occurs anywhere near you, it is well worth your effort to see it.

Partial solar eclipses, where the Sun is only partially covered by the Moon, are more common, and are also well worth observing. Be very careful when observing a solar eclipse, however, because even though the Sun is partially covered by the Moon, it is still bright enough to severely damage your eyesight if you observe without proper protection.

ALERT

It is safe to observe a total solar eclipse only with special glasses or a filter except during totality, a period of a few minutes at the height of an eclipse when the Sun is completely covered.

A lunar eclipse occurs when Earth's shadow covers the Moon. Lunar eclipses are more common than solar eclipses, can be seen over larger portions of the world, and don't require any special eye protection. During a lunar eclipse, which can also be partial or total, Earth's shadow begins to cover up the full Moon, eventually darkening its entire face. During the peak of a total lunar eclipse, you may be able to see the darkened disk of the Moon take on a reddish or other color. This color depends on the types of dust and gas currently present in Earth's atmosphere.

CHAPTER 16

Ground-Based and Space-Based Observation

For amateur astronomers, stargazing is something they do for entertainment, not their livelihood. Professional astronomers, however, require much advanced training—usually a doctorate in astronomy or physics. Then, after many years of school and practical training, a professional astronomer is finally able to use the best astronomical toys in the world, including huge telescopes that can see deep into the universe.

Overview

Professional astronomers make a serious commitment to the tools of their trade, and often help design instruments for new and improved observations in their specialized areas. Optical telescopes and radio telescopes are the two major types installed on Earth. The main difference between them is that they operate at different wavelengths, and can therefore detect different energies of electromagnetic waves. In other words, when they look at the same object, they see different things. Each object in the sky gives out a variety of radiation, so both types of telescopes are necessary for us to have a complete understanding of any given object.

Optical Versus Radio

Because of their wavelength sensitivities, radio and optical telescopes look very different. Optical telescopes look like a scaled-up version of the kind of telescope you might have in your back yard. Most often, an optical telescope is housed in a dome, and has a big mirror inside that reflects light to a detector—occasionally an eyepiece, but more often a computer chip for viewing on a computer monitor.

SSENTIALS

Solar telescopes are special optical telescopes used for observing the Sun. They require filters to keep them from getting too hot. Because they are so specialized, it is common for many observatories to have optical, radio, and solar telescopes.

A radio telescope, on the other hand, looks like an enlarged version of a back yard television satellite dish. Radio telescopes are much bigger than optical telescopes because the wavelength of light at which they operate is much longer. Radio telescopes don't even need to be one unit—the Very Large Array (VLA) in New Mexico is made up of a number of radio dishes that operate together as one enormous telescope.

Different Emissions, Different Technologies

To understand why astronomers use different telescopes for different observations, one needs to understand the type of energy that stars emit. Think of electromagnetic radiation as energy. Waves of light at different wavelengths are emitted by anything that gives off energy; the shorter the wavelength, the higher the energy. The Sun, for example, gives out energy in a huge range of wavelengths, but only a small portion of these colors can be seen—better known as the visible spectrum. Purple (violet) is the shortest wavelength humans can see, and red is the longest.

Wavelengths of radiation that are too short (too high energy) to be in the visible spectrum include ultraviolet, x-ray, and gamma ray. Some wavelengths are too long to be seen, meaning they are longer than red (infrared). Heat coming off a cup of steaming coffee is infrared radiation. Microwaves and radio waves are even longer than infrared.

Astronomy involves all these wavelengths. Specialized telescopes and instruments can detect wavelengths from radio waves to x-rays and gamma rays. Much of what we learn about stars requires telescopes that can observe wavelengths that are invisible to the human eye.

FIGURE 16-1:

The electro-magnetic spectrum

(refer to page 280 for more information)

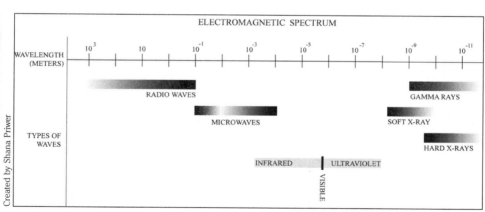

When we want to detect energy at wavelengths longer than infrared, we need to observe microwaves and radio waves. Since these wavelengths are too long to detect with a normal telescope, astronomers use radio telescopes at this point. The longer the wavelength you're trying to detect, the bigger the telescope you need. Microwave telescopes can look like a good-sized back yard satellite dish, but radio telescopes must have huge dishes to detect waves that are many meters in wavelength.

At the shorter end of the spectrum, past the ultraviolet, are x-rays and gamma rays. To study x-ray and gamma ray energy from stars, space-based telescopes such as Chandra and Compton are necessary because our atmosphere blocks most of these waves. Although it hampers some research, the atmospheric absorption is fortunate for life on Earth—high-energy radiation could be very dangerous if it reached the ground unimpeded!

Ground-Based Observation at the National Optical Astronomy Observatories

Large optical telescopes are responsible for much of the information we have about the universe, but different telescopes have different capabilities and can yield complementary data. The main restriction on ground-based telescopes is that we are limited in what we can see. The atmosphere is in constant motion, so most images received from ground-based telescopes appear somewhat fuzzy as they attempt to focus on objects far beyond the atmosphere. The logic behind deciding where to put a ground-based telescope, then, dictates that we place them as high and dry as possible.

The National Optical Astronomy Observatories (NOAO) is perhaps the largest organization for optical telescopes and observations from Earth. There are four major divisions in NOAO:

1. The Cerro Tololo Inter-American Observatory (CTIO) in Chile
2. Kitt Peak National Observatory (KPNO) in Arizona
3. The National Solar Observatory (NSO) in Arizona and New Mexico
4. NOAO's Science Operations division (SCOPE) in Arizona

The Cerro Tololo Inter-American Observatory is a group of telescopes and other astronomical instruments located in Chile. There are currently six telescopes available for observing by professional astronomers from around the United States. Among them is a RodoDIMM, a sort of seeing monitor. The DIMM (Differential Image Motion Monitor) is a special monitor attached to a telescope (in this case a 10 inch Meade) that can observe by itself every night. The DIMM comes in very handy for those nights when scientists want results without frostbite.

Cerro Tololo also has the Southern Gemini Observatory (the northern component resides in Hawaii); these two telescopes work together to provide simultaneous and sequential observations of the same object from two locations.

FIGURE 16-2:
Radio and optical telescopes on Kitt Peak, near Tucson

(refer to page 280 for more information)

© 2001, www.arttoday.com

The Kitt Peak National Observatory, located in southern Arizona, also has different types of telescopes for professional astronomers. KPNO has a 3.5-meter WIYN telescope, which was one of the first telescopes to use adaptive optics, meaning the telescope adjusts the shape of the mirror to account for distortions caused by the atmosphere. This telescope gives the viewer the sharpest possible picture, which is an incredible aid to professional astronomers.

FACTS

Other observatories in the United States include Mauna Kea in Hawaii, Whipple in Arizona, and Lick and Mount Palomar, both in California. These sites were all originally far from major cities. However, in the case of both Lick and Palomar, the cities (and their ambient light) are rapidly reducing the quality of these sites' observations.

The National Solar Observatory is located in two sites: the NSO at Sacramento Peak in New Mexico, and the NSO at Kitt Peak in Arizona. The McMath Solar Telescope at the Kitt Peak site is one of the world's largest solar telescopes. There is also a solar vacuum telescope at Kitt Peak, used for observing magnetic fields, solar activity, and solar forecasting.

The NSO at Sacramento Peak consists of three main parts. The Dunn Solar Telescope is famous for high-resolution images of the Sun. The Evans Solar Facility uses several telescopes to observe the solar corona, solar flares, and other exciting phenomena. Finally, the Hilltop Dome Facility has a computer-controlled guidance system to always keep the telescopes focused properly on the Sun.

The fourth component, SCOPE, is a division of the National Optical Astronomy Observatories that coordinates observing trips to the telescopes and services that NOAO administers.

Radio Observatories

As the name would suggest, radio astronomy is the study of celestial objects by the radio waves they emit. Radio astronomy began, in a way, with the Guglielmo Marconi's first radio transmission in 1901. Amateurs were given the short wavelengths to listen to (termed *shortwave radio*), and the first transatlantic reception occurred in 1921. Shortwave communication aroused great interest by telephone companies, and engineers from AT&T studied the static that such communications produced. This static was later identified as coming from other parts of the galaxy. In the 1930s Grote Reber built one of the first satellite dish antennas, about 32 feet in diameter, and larger ones continue to be built today.

FIGURE 16-3:
The dome covering a radio telescope at Kitt Peak, near Tucson

(refer to page 280 for more information)

© 2001, *www.arttoday.com*

One problem with radio astronomy is interference. Some signals from the cosmos are so weak that even when astronomers—with the best equipment—are listening on certain frequencies, radio stations transmitting on nearby frequencies can cause interference. Static can even be caused by poorly designed devices we use every day, such as garage-door openers and cellular telephones, as well as by transmitters that orbit Earth in satellites.

The National Radio Astronomy Observatory is part of the National Science Foundation, and builds radio telescopes for use all around the world. Radio telescopes are ideal for studying pulsars, quasars, and hypothesizing about radiation from the Big Bang. In fact, they are virtually the only way to study these types of subjects, and in this respect radio astronomy has revolutionized our ability to gain knowledge about the cosmos.

E

FACTS

The largest radio telescope in the world is located in Arecibo, Puerto Rico. Part of the Arecibo Observatory, this gigantic dish is 1,000 feet in diameter and 167 feet deep, and its surface is made up of 40,000 aluminum panels.

FIGURE 16-4:
Arecibo radio telescope

(refer to page 280 for more information)

Courtesy of Taylor Bucci, 2001

The Very Large Array (VLA) in New Mexico, consists of twenty-eight separate dish antennas. The VLA was constructed between 1973 and 1981. Each dish has a diameter of 81 feet, with a combined antenna amounting to 422 feet in diameter. The dishes are laid out in a Y-pattern spread over an area that is larger than Washington, D.C. The telescopes can be moved into different configurations, depending on their observing target.

Solar Telescopes

Solar telescopes study the Sun and its atmosphere. As the primary source of energy here on Earth, the Sun is an important subject for research. We are able to study surface elements of the Sun better than we can the surface of any other star, so it's one of our main sources of information concerning the physical and chemical makeup of stars. In addition to the National Solar Observatories mentioned earlier, there are a number of other solar telescopes in the United States, including the Mount Wilson observatory, located in southern California, and the Big Bear Solar Observatory (BBSO), also in California.

The Advanced Technology Solar Telescope (ATST) is a proposal in the planning stages for what would be a truly enormous solar telescope. The design documents call for a 4 meter aperture and unprecedented optics. If constructed, it could provide amazing additions to our knowledge of solar science. The ATST is currently being studied by a group of twenty-two different institutions interested in solar research.

Space-Based Observation

Telescopes on the ground can observe planets, galaxies, and astronomical phenomena, but Earth's atmosphere restricts both wavelength and resolution by blurring images and absorbing certain wavelengths. Telescopes in space resolve this problem by moving out beyond the atmosphere.

Getting telescopes into space posed a major technical challenge. Mounting the telescopes onto rockets wouldn't work because rockets were in space for only a few minutes before falling back to Earth. Mounting them on weather balloons and airplanes was another option, but neither was able to completely get beyond our atmosphere. That left mounting the telescopes on orbiting satellites.

In order to take continuous observations, space-based telescopes need to be in a stable orbit around Earth. NASA has created an ambitious series of Great Observatories—telescopes in orbit around Earth that can observe at various wavelengths. Three of these telescopes have already been launched, and the fourth is scheduled for launch in mid-2002. These telescopes have already revolutionized our views of the cosmos.

Hubble

The Hubble Space Telescope, carried into orbit in 1990, was the first dedicated space-based observing telescope, and the first component of NASA's Great Observatories program. The mirror measurement was constrained by the size of the cargo bay of the space shuttle, but the information it has been able to send back has shed light on planets and events of which we would otherwise have minimal knowledge. Hubble has sent back pictures from many exciting objects, including star clusters, warped galaxies, and amazing pictures of the planets.

Compton

The Compton Gamma-Ray Observatory was launched by the space shuttle in 1991. It collected data on high-energy processes in the universe, including some extremely violent events, for almost ten years. The end of Compton's mission was reached in mid-2000, and it safely re-entered Earth's atmosphere and landed in pieces in the Pacific Ocean on June 4, 2000.

Chandra

The Chandra X-ray Observatory is NASA's major x-ray telescope. Launched in 1999, Chandra can observe and record x-ray data from exploded stars such as the Crab Nebula. The amount of detail Chandra's image spectrometer can pick up helps us to understand the connection between pulsars and energy, among other things. Chandra is about 45 feet long, and is the largest satellite ever launched from the space shuttle *Columbia*. The observatory consists of the spacecraft that houses the equipment, an x-ray telescope, and recording instruments.

FACTS

The Space Infrared Telescope Facility (SIRTF), scheduled for launch in July 2002, will be the fourth and last of NASA's Great Observatories. SIRTF will study processes occurring in the universe in the thermal infrared. Infrared observations can study the remnant background radiation from the early days of the universe.

Chandra has a built-in booster, which allows it to orbit about 200 times higher than Hubble, and gives it a much wider field of view. All space-based telescopes are operated remotely, so the utmost precision goes into their design. Chandra enables astronomers to study hot regions, dark matter, supernovas, and many other cosmic features in ways we've never done before. The images taken aboard the satellite return to the Chandra X-ray Center (CXC) in Cambridge, Massachusetts, for processing and analysis.

Looking Ahead

In the future, what other advances might be made in space-based telescopes? One possibility is the construction of a telescope on the Moon. A lunar telescope has many advantages over Earth-based telescopes. The lunar surface is extremely dark, without the light pollution that plagues terrestrial observatories. There is also almost no atmosphere, so a telescope on the Moon would have the same crystal-clear view available from Earth's orbit, but without the complexities of having a telescope located on an orbiting satellite.

A telescope could also be located on the far side of the Moon (the "dark" side), which permanently faces away from Earth. This location could be especially good for a large radio telescope, since the Moon's mass would block out most of the radio interference from terrestrial sources.

SSENTIALS

At one-sixth Earth's gravity, the Moon could host a large telescope with a minimum of building material. However, the Moon has at least *some* gravity, making assembly easier than building in a zero-gravity space environment.

NASA has no current plans to build such a telescope, but once the International Space Station is fully constructed and operational, it seems likely that NASA will turn to the Moon for its next big exploration program. If NASA is considering building a base on the Moon for exploration anyway, what better purpose to use it for than astronomy? Perhaps in the next few decades, astronomers all over the world will have a chance to perform observations of unprecedented quality.

CHAPTER 17

Robots, Animals, and People in Space

The Space Age began with the success of the Soviet satellite *Sputnik* in 1957, the first human-made object to orbit Earth. Since this auspicious launch, many robotic spacecraft have begun the exploration of the solar system. However, there are certain things that just cannot be accomplished by even the most sophisticated robot, so once the preliminary reconnaissance is completed, crewed missions are the necessary next step.

Early Missions

Robotic exploration is a relatively fast and easy way to explore our solar system, and we can also send robots into environments where humans would have a difficult time surviving, such as regions of high radiation. As we learn more about what awaits us in each phase of exploration, we'll have a better understanding of humans' roles in that exploration. It takes research and patience, as it has all along.

The Moon

The first robotic missions to another object were the Soviet Luna missions. *Luna 2* crash-landed on the Moon in 1959, and *Luna 3* took the first photos of the far side of the Moon.

Following these early events, a series of robotic missions from both the United States and former Soviet Union studied the Moon in depth before the NASA Apollo missions carried astronauts to the lunar surface. The U.S. Ranger missions crashed into the Moon, taking pictures on the way down. In 1966 and 1967, five NASA Lunar Orbiter missions produced detailed maps of the Moon's surface to help select *Apollo* landing sites. These missions also provided the first detailed maps of the far side of the Moon.

QUESTIONS?

Which planets have had landers placed on their surfaces?
Venus (the Soviet *Venera* landers), Mars (U.S. *Viking* and *Pathfinder*), the Moon (U.S. *Surveyor*, Soviet *Lunokhod*, U.S. *Apollo* manned landers), the asteroid Eros (NEAR), and Jupiter (the United States sent an entry probe, which unfortunately found no solid surface to land on).

The thousands of pictures of the Moon taken by the *Lunar Orbiters* have never been put into digital form, but if digitized, they would yield more data than any other planetary mission to date. The five successful Lunar Surveyor missions were a series of NASA missions from 1966 to 1968 to practice making a soft landing on the surface of the Moon; they also made sure that the dusty lunar surface could support a spacecraft. The *Apollo 12* spacecraft ended up landing only 600 feet from the *Surveyor 3* spacecraft in 1969, and parts of the spacecraft were returned to Earth for analysis.

A series of Soviet missions also studied the Moon, although plans for a crewed landing were abandoned due to two factors: a series of failures in the Soviet program, and the fact that the United States managed to land people on the Moon first (removing the political agenda in the Cold War era).

Venus and Mercury

The U.S. *Mariner 2* mission performed the first successful flyby of Venus in 1962, and confirmed the very hot surface temperature of that planet. Following *Mariner 2*, the Soviet Union launched a series of highly successful Venus missions. *Venera 9* successfully landed and sent the first image of the surface of Venus back to Earth in 1975, and was also the first mission to orbit Venus. Color images were returned from the surface of Venus by *Venera 13* and *14* in 1982, and radar mapping of the surface was performed in 1983 by *Venera 15* and *16*.

FIGURE 17-1:
Venera 13,
a Venus
lander built
by the former
Soviet Union

(refer to page 280 for more information)

Courtesy of National Space Science Data Center (NSSDC)/NASA

Venus was also studied by the United States, including the *Mariner 5* spacecraft in 1967. *Mariner 5* measured the atmospheric pressure at the surface of Venus as being at least ninety times the surface pressure on Earth. *Mariner 10*, in 1974, took detailed images of the clouds of Venus before continuing on to Mercury. It is still the only probe to have taken images of the planet closest to the Sun, and revealed Mercury to be a planet covered mostly with craters, with a very high mass due to a large iron core.

Mars

Mariner 4 took the first close-up pictures of the surface of Mars in 1965, and saw mostly the southern, heavily cratered hemisphere. These photos, along with the knowledge that the martian astmosphere is very thin, convinced scientists that Mars is an old, geologically dead world like the Moon. *Mariners 6* and *7* in 1969 also supported the view that Mars was an old, cratered body.

Mariner 9, the first Mars orbiter in 1971, revealed a much more geologically interesting planet than the previous flybys, and also took images of the two small moons of Mars, Phobos and Deimos. *Mariner 9* showed that the Red Planet was in fact a diverse world with volcanoes, canyons, and the possibility of liquid water in the past. *Viking 1* and *2* reached Mars in 1976, and performed a search for life. This search had inconclusive results, but most scientists believe it did not detect life, at least the kind of life it was programmed to search for.

Outer Solar System

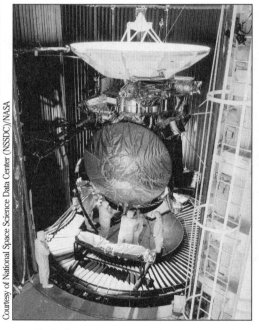

FIGURE 17-2:
The *Cassini* space probe undergoes preflight testing

(refer to page 280 for more information)

Courtesy of National Space Science Data Center (NSSDC)/NASA

The first missions to the outer solar system were *Pioneer 10* and *11*. *Pioneer 10* was the first spacecraft to fly by Jupiter in 1973, and *Pioneer 11* flew past Jupiter in 1974 and then past Saturn in 1979. The spacecraft had mainly fields and particles instruments, and took only rudimentary pictures of Jupiter and Saturn. Both probes are heading out into interstellar space, and *Pioneer 10* is still operating and sending occasional data back to Earth!

Voyager 1 and *2* followed the lead of the *Pioneers*, and these immensely successful missions studied the planets of the outer solar system in detail for the first time. Both were launched in 1977, and flew by Jupiter in 1979. They discovered three new satellites, the rings of Jupiter, and characterized the complex dynamics of Jupiter's atmosphere. They also observed lightning and auroras on Jupiter.

Voyager 1 reached Saturn in 1980, and *Voyager 2* arrived there in 1981. At Saturn, the two probes discovered seven satellites and mapped the complex ring system, including many discontinuous ring arcs. The atmosphere of Saturn was found to be much less dynamic than that of Jupiter. Titan, Saturn's largest satellite, remained an enigma due to its thick smoggy atmosphere.

FACTS

The two *Voyager* spacecraft are still operating, and are expected to reach the heliopause (the end of the Sun's influence on interstellar space) soon.

Voyager 2 took advantage of a planetary alignment occurring once every 175 years to continue on a grand tour of the outer solar system. It reached Uranus in 1986, and found that its magnetic axis was even further offset from the rotation axis. Ten new satellites were discovered, and images of those icy satellites were taken. Miranda in particular was shown to have a peculiar jigsawlike appearance, as if the whole satellite had been taken apart and put back together haphazardly.

Voyager 2 continued on to Neptune in 1989. Neptune had much more interesting weather patterns than Uranus, and six new satellites and two new rings were discovered. Particularly interesting was Neptune's large moon Triton, which had a very young surface. Ongoing geologic activity in the form of geysers was observed on Triton, and much of the surface was covered with a peculiar mottled, cantaloupe-textured terrain.

Modern Missions

The 1990s saw a new wave of planetary exploration from a variety of probes, mainly from NASA but also from other countries. These probes were able to explore many of the planets of the solar system in detail. The following is a list of bodies specifically targeted by probes launched since 1989 by the United States (and one launched by Japan):

VENUS: *Magellan*—*Magellan* was launched in 1989 and performed radar mapping of 98 percent of the surface of Venus beginning in 1990. It also collected detailed information on the internal gravity field of Venus and tested the theory of aerobraking (using the atmosphere of Venus to slow and circularize *Magellan's* orbit). Aerobraking has since been used on Mars missions, and this technique has resulted in saving significant amounts of fuel.

JUPITER: *Galileo*—Also launched in 1989, the *Galileo* spacecraft took a circuitous route that involved two Earth flybys and one Venus flyby just to gain enough energy to send it out to Jupiter, which it reached in 1995. *Galileo* made direct observations of the impacts of the pieces of comet Shoemaker-Levy 9 into Jupiter, and made many observations of Jupiter and its satellites.

SUN: *Ulysses*—Launched in 1990, this spacecraft's main goal was to study the solar wind and interplanetary magnetic field, as well as to record energetic particle compositions and accelerations. *Ulysses* was a joint project of ESA (the European Space Agency) and NASA, and is still studying the Sun.

ESSENTIALS

The first space shuttle flight took place in 1981. The Space Transportation System (STS), more readily known as the space shuttle, ushered in an age of reusable spacecraft, eventually making access to Earth's orbit almost routine.

MOON: *Clementine*—*Clementine*, launched in 1994, was a joint mission funded by the Strategic Defense Initiative Organization (part of the Department of Defense) and NASA. Its primary mission was to test a variety of sensors and other spacecraft components developed for military satellites, but under the auspices of NASA it was able to perform some important science as well.

ASTEROIDS: *NEAR*—NEAR (Near-Earth Asteroid Rendezvous) was launched in 1996 as the first of NASA's "Faster, Better, Cheaper"

Discovery-class missions. In 2000, NEAR became the first spacecraft to orbit an asteroid, and it took thousands of detailed pictures of the surface of the asteroid Eros. After spending a year mapping Eros in great detail, scientists decided to try to make *NEAR* the first spacecraft to land on an asteroid—it eventually touched down softly on the surface of Eros, where it spent two weeks transmitting data!

MARS: *Pathfinder* and *Global Surveyor*—Both probes were launched in 1996 and arrived at Mars in 1997. Mars *Pathfinder* included a stereo imaging camera, a weather experiment, and a small rover named Sojourner. The rover rolled down a ramp onto the surface, and performed a variety of experiments on the soil and various rocks at the landing site.

SATURN: *Cassini*—The *Cassini* mission was launched in 1997, and is currently en route to Saturn, where it will arrive in 2004. There, it will deploy an atmospheric probe built by ESA called *Huygens*, which will descend through the atmosphere of Saturn's large moon Titan. Titan's thick atmosphere makes it difficult to see the surface, so if the probe survives the trip to the ground it may provide our first pictures of the surface of Titan.

MOON: *Lunar Prospector*—Another Discovery mission called *Lunar Prospector*, launched in 1998, performed a 1.5-year-long mission mapping the surface composition of the Moon. *Lunar Prospector* also looked for any evidence of polar ice deposits, hints of which had been detected from Earth-based radar observations and from the Clementine spacecraft. The evidence for such water ice remains inconclusive.

MARS: *Nozomi*—The first interplanetary Japanese spacecraft was launched in 1998. It was originally scheduled to reach Mars in 1999, but its arrival has been delayed to 2003. *Nozomi* will mostly study the upper atmosphere of Mars and how it interacts with the solar wind, but will also take some images of its surface.

COMETS: *Stardust*—The *Stardust* spacecraft is a NASA Discovery class mission, launched in 1999. It will fly by comet P/Wild 2 in 2004 and take samples of the dust and volatiles in the comet's coma, as well as samples of interstellar dust particles, and return them to Earth. It will also take close-up images of the nucleus of comet P/Wild 2. After collection, the samples will be placed in a sealed re-entry capsule, and sent back to Earth in 2006.

MARS: 2001 *Mars Odyssey*—The 2001 *Mars Odyssey* spacecraft, the next in the series of Mars exploration spacecraft, was launched in April 2001 and arrived at Mars in October 2001. The mission's main goals are to help understand whether the surface of Mars was ever able to support life. It will study the climate and geology of Mars, and examine the radiation environment to understand what risks might be in store for future astronauts on Mars.

SOLAR WIND: *Genesis*—The *Genesis* spacecraft, launched in 2001, is another NASA Discovery spacecraft that will collect samples of particles from the solar wind and return them to Earth for study. The samples obtained will help scientists better understand the formation and evolution of the solar system during its early days.

Early Space Travelers

Sending astronauts into space is one of humankind's greatest achievements. The ability to see Earth without any political boundaries has given us an entirely new perspective on Earth as a planet— scientifically, socially, and emotionally. The twelve U.S. astronauts who have walked on the surface of the Moon had the opportunity to see firsthand what living on another planet might be like, generating entirely new perspectives on colonization of other planets.

FACTS

The first living creature in space was a Russian dog named Laika. She traveled on *Sputnik 2* on November 3, 1957. The second animal to experience the cosmos was Gordo, a monkey from the United States aboard the *Jupiter AM-13* on December 13, 1958.

The first person to travel in space was Yuri Gagarin (1934–1968). After becoming a member of the Soviet Air Force in 1955, Gagarin joined the Soviet cosmonaut training program. He first orbited Earth on April 12, 1961 in the *Vostok 1*. Lasting 108 minutes, his space flight was launched from the Baikonur cosmodrome in the Soviet Union, orbiting Earth at speeds up to 27,400 kilometers per hour. *Vostok 1* used shortwave and VHS communications to remain in contact with the Soviet base. Gagarin ejected by parachute, and returned safely to Earth.

Alan Shephard (1923–1998) was the first American to fly into space. On May 5, 1961, the Naval Academy graduate flew a fifteen-minute suborbital flight in the *Freedom 7* Mercury capsule. Although he was considered for a Gemini mission, his next space flight was not until *Apollo 14*, January 1971. He became a founding member of the National Space Society (NSS) in the late 1970s. The NSS is an educational group dedicated to the exploration of space civilization.

ALERT

Space travel has its dangerous side. The worst space disaster in American history was the explosion of the space shuttle *Challenger* minutes after takeoff from Florida. All seven astronauts on board were killed, including the first teacher in space, Christa McAuliffe.

The first American to orbit Earth was John Glenn (1921–). Glenn served as a U.S. Marine Corps pilot in World War II and the Korean War. In 1959, Glenn was one of seven people chosen for the Mercury program. Glenn orbited Earth three times in the *Friendship 7* on February 20, 1962. Total flight time was four hours, fifty-five minutes. The flight took him across the Atlantic, then over Africa, the Indian Ocean, and Australia. His contributions did much to alleviate the fear that the Soviets were the only ones making advances in space and aeronautics, and helped the American space program gain popularity.

Glenn is, coincidentally, also the oldest American to travel in space. On October 30, 1998, at the age of seventy-seven, John Glenn took part in a nine-day mission on the space shuttle *Discovery* (STS-95) to participate in biological tests, including the effects of space on the aging process. Glenn was a designated payload specialist on the mission; his duties included running experiments in the zero-gravity environment of space. The training and analysis Glenn endured before, during, and after the mission were intense; fluid samples, monitoring devices, and other physical tests were administered constantly. Glenn's flight attracted international publicity because of his celebrity status; his role not only provided scientific information, but also revitalized interest in the space shuttle program.

ESSENTIALS

In April 2001 Dennis Tito became the first "space tourist." He paid the Russian government approximately $20 million to travel to the International Space Station. Since his successful flight, Tito is investigating the possibility of future space tourism.

United States Astronaut Missions

Project Mercury was the first crewed series of U.S. space missions. Taking place between 1961 and 1963, the six missions aimed to orbit humans around Earth, explore a human's ability to survive in space, and successfully return both the spacecraft and its occupants to Earth. Alan Shepherd's Mercury *Freedom 7* mission impressed President Kennedy to such an extent that he immediately pronounced his goal of landing an American on the Moon before the end of the 1960s. The Mercury missions included the *Freedom 7*, *Liberty Bell 7*, *Friendship 7*, *Aurora 7*, *Sigma 7*, and *Faith 7*.

ESSENTIALS

There were seven Mercury astronauts: Alan Shepard, Virgil "Gus" Grissom, L. Gordon Cooper, Walter Schirra, Donald "Deke" Slayton, Scott Carpenter, and John Glenn.

Project Gemini consisted of ten crewed spaceflights between 1962 and 1966. This second series of missions set out to test the duration of humans and equipment in space, to dock with other spacecraft currently in orbit, to further our knowledge of entering and leaving Earth's atmosphere, and to attempt controlled landings on Earth's surface.

Gemini spacecraft improved upon the Mercury designs, and the planned dockings allowed for experimentation and consideration of how a Moon landing might work.

FIGURE 17-3:
Edwin Aldrin lands on the Moon

(refer to page 280 for more information)

Courtesy of NASA/Johnson Space Center [JSC]

Project Apollo was the United States's third crewed series of spaceflight missions. The objectives were clear: land a human on the surface of the Moon, explore the Moon's surface for as much useful information as possible, devise the technology to allow humans to work in a lunar environment, and create understanding for promoting the role of the United States as a pre-eminent force in the space race.

Lunar Landings

Six Moon landings were accomplished over the course of eleven manned launches (with a total of seventeen unmanned launches).

FIGURE 17-4:
A footprint on the surface of the Moon

(refer to page 280 for more information)

Courtesy of NASA/Johnson Space Center [JSC]

The Apollo missions involved two modules—a command module (CM), and a lunar module (LM). While one crew member orbited the Moon inside the command module, the other two astronauts took the lunar module down to the Moon's surface to take pictures, collect samples, and then return to the command module for the trip home to Earth.

There were, of course, some trials prior to the first successful moonwalk. *Apollo 1* caught fire on the launch pad during a preflight test and all three astronauts on board (Virgil Grissom, Ed White, and Roger Chaffee) were killed; this accident was the only fatality in the U.S. space program to that point. Neil Armstrong became the first American to set foot on the Moon with *Apollo 11* (Columbia CM, Eagle LM) in 1969, followed shortly thereafter by Edwin "Buzz" Aldrin; their mission was tracked from orbit by the third astronaut, Michael Collins.

Apollo 13 was the only mission between *Apollo 11* and *Apollo 17* that did not land on the Moon. Astronauts James Lovell, John Swigert, and Fred Haise were supposed to land in the Fra Mauro region of the Moon but an explosion took place on board, leading to the infamous understatement "Houston, we have a problem." The explosion required the crew to orbit the Moon instead of landing. Thanks to the ingenuity of the crew and a large number of NASA engineers on the ground, the astronauts returned to Earth safe and sound.

The Space Shuttle

The space shuttle program has provided reliable transportation into space for the last twenty years. Space shuttle missions have taken astronauts, satellites, and a variety of scientific experiments into orbit. The space shuttle consists of three basic components—an orbiter with engines, rocket boosters, and an external fuel tank. Only the orbiter and the engines actually go into orbit around Earth; the rocket boosters and fuel tank drop off after launch. The major fuel tanks are reusable, and are picked up from the ocean and refilled for later use. There have been five space shuttles to date—*Columbia, Discovery, Challenger, Atlantis,* and *Endeavor,* as well as the test model, Enterprise, which never flew into space.

FIGURE 17-5:

The space shuttle *Columbia*

(refer to page 280 for more information)

Courtesy of National Space Science Data Center (NSSDC)/NASA

Columbia is the oldest space shuttle in the fleet. First launched in 1981, it stands 37.2 meters tall, with a maximum width of 23.8 meters. The shuttle can accommodate up to eight crewmembers, with over 70 cubic meters of inhabitable space. Columbia had successfully completed twenty-six spaceflights at the start of the year 2000.

The second space shuttle, *Challenger,* was designed as a structural test orbiter, but was upgraded to a space-quality shuttle by 1982. *Challenger* had nine successful missions, but was destroyed in a tragic accident during its tenth flight in 1986. Unexpectedly cold temperatures caused mechanical failure following launch, and all the crew members died. Their names were Gregory Jarvis, Ronald McNair, Ellison Onizuka, Francis Scobee, Judy Resnik, Michael Smith, and S. Christa McAuliffe. All shuttle missions were suspended for two years following this disaster.

In use since 1984, the third shuttle, *Discovery*, has retrieved satellites, deployed the Hubble Space Telescope in 1990, rendezvoused with the Russian Mir Space Station in 1995, and allowed John Glenn to return for another trip into space in 1998.

Completed in 1984, space shuttle *Atlantis* was the fourth shuttle built. It weighs nearly 7,000 pounds less than *Columbia*, and was quicker to construct. It has been on missions since 1986, including the launching of the *Galileo* spacecraft and the *Magellan* probe.

If you want to watch a space shuttle launch, make sure you are in an approved viewing area, not in a boat or plane in the danger zone. People in the wrong places at the wrong times have caused postponements of shuttle launches.

Endeavor was built as a replacement following the *Challenger* accident, and was completed in 1991. *Endeavor* has assisted in the construction of the International Space Station, as well as many other missions.

The International Space Station

The International Space Station, or ISS, was in the planning stages for many years. The space station was originally a United States–only project, but budget cuts and improved United States–Russia relations made it sensible (if not necessary) to collaborate with as many countries as possible to fulfill this dream. Now, the ISS is a joint project of fifteen countries and five space agencies: the Canadian Space Agency, the eleven members of the European Space Agency, Japan's NASDA (National Space Development Agency), Russia's space agency, and the United States's NASA. Each country and space agency brings not only funding, but also experience in different areas related to the building of the space station.

The first module of the space station was launched in late 1998. To date, there have been many delays in the station's construction, and repair missions have begun although the space station itself is not complete. Ultimately, over forty space flights will complete the building of the space station; its components will be delivered by the space shuttle, the Russian *Soyuz* rocket, and the Russian *Proton* rocket.

FIGURE 17-6:
Artist's conception of the finished International Space Station

(refer to page 280 for more information)

Courtesy of National Space Science Data Center (NSSDC)/NASA

Only a few dozen people have had the privilege of seeing Earth from orbit, but space travel has the possibility of becoming more commercially accessible in the future. The space program today is run by a government agency, a direct result of the Cold War competition with the Soviet Union and other countries. As the private sector develops more aeronautic technology, including reusable launch vehicles, there is an increased possibility that people may one day be able to purchase a space shuttle ticket and take a ride around Earth.

Spacewalks

Spacewalks, or EVAs (extravehicular activity), are one of the most exciting aspects of space flight for astronauts and civilians alike. Ed White, an astronaut with the Gemini missions, was the first American to walk in space in 1965. An umbilical cord supplied his spacesuit with oxygen and power, and this first EVA lasted just over twenty minutes. With current technology, spacewalks can last for several hours.

FACTS

The longest spacewalk to date is eight hours, fifty-six minutes. Astronauts Jim Voss and Susan Helms set this record in March 2001 while working on the International Space Station.

Of course, astronauts won't survive the walk without space-age armor: spacesuits. Not only are the suits pressurized to keep body fluids from boiling or freezing, they have a range of other functions. They provide air pressure and oxygen, eliminate carbon dioxide, protect from flying rocks or space debris, keep a comfortable temperature, allow for fog-free vision, and accommodate regular movement and communication. Oxygen is pumped in and carbon dioxide is pushed out, so that the air the astronauts breathe is as close as possible to that on Earth. The helmets are usually tinted to ward off the effects of solar flares, and are sprayed with an antifog substance. Spacesuits are temperature controlled and equipped with radio communications devices.

CHAPTER 18

SETI: Are We Alone?

The search for life elsewhere in the universe has captured our imaginations ever since we discovered there were other stars and planets out there. Early astronomers imagined that a great civilization had built extensive canals on Mars, and that Venus's thick atmosphere shrouded a jungle world full of fantastic creatures. We know a little more now than we did then, but how much? Read on and find out.

Habitable Zones

The search for other life in the universe begins with the search for other solar systems. Our definition of habitable planets describes Earth-like planets. But in a way, this bias is natural—the only example we have of a planet with life is Earth, and the only life we know of is terrestrial.

FIGURE 18-1:
Habitable zones

(refer to page 280 for more information)

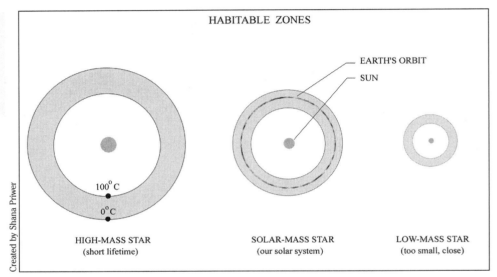

Created by Shana Priwer

Life as we know it is based on water, the universal solvent. Water is a unique substance, and makes up a large percentage of every living creature on Earth. Water is also present on Earth in all three forms—solid ice, liquid water, and gaseous water vapor. If we expect that life elsewhere would also be based on liquid water, then we can define a habitable zone surrounding a star as the distance from a star where liquid water can be present on the surface of a planet. Large stars will have a large habitable zone, and small stars a small habitable zone.

Large Star, More Habitable?

If larger stars have larger habitable zones, are they the best places to look for life? Not necessarily. As scientists understand it, life takes quite a while to develop on a planet, especially intelligent life. But remember that the larger the star, the shorter its lifetime. Large stars will likely have too short a lifetime to allow stable conditions on their planets to persist long enough for intelligent life to come about. This situation holds true even if they do have a large habitable zone with many planets in it, since life most likely develops independently on each planet (with perhaps some transfer back and forth via meteorites). If we assume that intelligent life takes about 3 billion years to develop on a planet, only stars with a minimum lifetime of 3 billion years are suitable. Such stars would end up having masses of 1.5 solar masses or less.

Small Star, Less Habitable?

There's also a problem if the star is too small. Less massive stars last much longer than big ones, but their habitable zones get much closer to the star since they give out much less energy. For a 0.5–solar-mass star, for example, the habitable zone is so close to the star that any planet in range probably would end up tidally locked to the star, with one side always facing the star and one side always facing away. The side facing the star would be extremely hot, and the side facing away extremely cold—not conditions too conducive to life! The minimum size for a star is therefore at least 0.5 solar masses. We now have a mass range of between 0.5 and 1.5 solar masses. Conveniently for us, our Sun is right in the middle!

Stellar Requirements for the Habitable Zone

Recall that some heavy elements are created in large stars, and even heavier ones must be created in supernovas. In order to have the necessary heavy elements for life, such as carbon, nitrogen, and iron, the star and planets must form from a gas cloud that is already enriched in these elements from previous stellar generations. Astronomers can actually detect metals in the spectra of faraway stars—if a star has metals in its spectrum, then any planets around that star probably will too. Therefore, stars that are more likely to have life on their planets can be identified.

For life as we know it to form, stars must be between 0.5 and 1.5 solar masses, sufficiently enriched in metals and other heavier elements, and in either single star systems or certain binary star systems. In other words, astronomers are looking for stars that are quite similar to our own Sun!

Planets must also be in stable orbits around their star. This requirement eliminates most complicated star systems with three or more stars orbiting each other, but stable orbits are feasible around closely separated binary star systems. Actually, if the stars in a binary (or greater) star system are separated widely enough, then planets that are relatively close to either star can have stable orbits as well. One example is the Alpha Centauri system, which happens to be the closest star system to our solar system. The distance at which the binary stars Alpha Centauri A and B orbit each other would allow planets close to either star to be in stable orbits. The Alpha Centauri system also contains a third star, Alpha Centauri C (or Proxima Centauri), but it is a small red dwarf star that is too small and cold to sustain habitable planets around it.

What Is Life?

Scientists still don't know for sure how life began here on Earth, so they certainly don't know how life would start anywhere else in the universe! In fact, it's very difficult to even come up with a universal definition of *life* that doesn't end up excluding things that are known to be alive, such as viruses, or including things that definitely aren't alive, such as fire or computer programs. Most scientists end up saying something like "we'll know it when we see it," which is difficult, if not impossible, when scientists are trying to detect life on other planets, either in our solar system, another solar system, or a different galaxy altogether!

Scientists agree on a basic list of characteristics that most life has in common:

1. **Life is organized,** and involves detailed structures at every level from microscopic to macroscopic.
2. **Life grows,** and uses energy to do so.
3. **Life responds to various stimuli,** either external or internal.
4. **Life reproduces,** and reproduction involves variation—mutations are introduced during the reproduction process, and favorable mutations are passed down to the next generation.
5. **Life adapts and changes** to better suit its environment through the mutation process, coupled with natural selection.

A more succinct definition of life is: a self-contained chemical system capable of undergoing Darwinian evolution.

None of these definitions of life are foolproof—if you look long and hard enough, it is almost guaranteed that you'll discover a living organism that doesn't follow one of the rules, or a nonliving construction that follows all of them. But these characteristics give scientists a starting point and guidelines for recognizing nonterrestrial life if we find it.

Life Elsewhere in the Solar System

In our solar system, only Venus and Earth are inside the habitable zone. The surface of Venus is much too hot, however, because of its thick atmosphere. But what about Mars, which is just outside the habitable zone? True, its surface is now cold and dry, and it has a very thin atmosphere. But what if Mars had a warmer, wetter past? Scientists believe such a previous period might have existed, and during this warmer time period Mars could have supported a thicker atmosphere and liquid water on its surface.

FACTS

The basic requirements for life to form are liquid water, certain biogenic elements (including carbon, nitrogen, oxygen, phosphorus, sulfur, and others), and a sufficient energy source.

Mars is an excellent place to look for signs of life, at least fossilized microscopic life. The first Martian lander in 1976 performed a variety of experiments to look for evidence of life on the surface of Mars. Basically, these experiments provided a food source, and looked to see if metabolic reactions took place. There was such a reaction, but it was later explained to be solely due to the very oxidized surface of Mars (the same reason why Mars is red), and not because there were actually Martian microorganisms performing chemical reactions. These results are somewhat controversial, so future Mars missions may perform more sophisticated searches for present or past life on Mars.

The problem with studying meteorites that have landed on Earth for signs of extraterrestrial life is that most of them have been sitting on the surface of Earth for thousands of years, and therefore most likely have suffered terrestrial contamination.

When astronauts finally make it to the surface of Mars, with luck sometime this century, they will be able to perform complicated onsite experiments. Geologist-astronauts on Mars will also be able to dig through cliffs and other outcroppings of buried rocks to look for evidence of fossil life that could have survived on Mars in the past.

Planetary Protection

One problem with sending people to Mars is the fear of terrestrial contamination. Just imagine if Mars does indeed have its own life forms. They are likely microscopic, but even so could have evolved completely separately from Earth life. If humans suddenly arrive on Mars, they will bring all sorts of terrestrial microscopic life forms with them. Such life forms might be able to colonize the planet and maybe even kill off any indigenous Martian life. It's also possible that any Martian life could be dangerous to humans and make us ill, though given the very different evolutionary paths, this course of events would be unlikely.

QUESTIONS?

What terrestrial life form could survive a trip to Mars on the outside of a spacecraft? with no atmosphere and extremely cold temperatures?

It seems impossible, but terrestrial microscopic organisms that can survive the rigors of space—no atmosphere, and extremely cold temperatures—for years at a time. A variety of bacteria orbited the Earth on the *Long Duration Exposure Facility (LDEF)* satellite and were exposed to space for more than five years; they were still able to grow and reproduce when they were returned to Earth.

FIGURE 18-2: Lunar astronauts in quarantine after returning from the moon

(refer to page 280 for more information)

Courtesy of National Space Science Data Center (NSSDC)/NASA

NASA scientists make sure we don't contaminate any other planets, and that we keep Earth safe from harmful non-Earth life forms (if any exist). When astronauts first returned to Earth from the Moon, they were rigorously decontaminated and kept in strict isolation until doctors were sure they didn't have any harmful space viruses or other diseases. After a careful examination of both the astronauts and their samples and equipment, NASA scientists decided that there were no harmful bacteria on the Moon, nor any kind of life at all, and subsequent missions didn't have to undergo the same kind of quarantine.

The same issues were considered when NASA was designing the first spacecraft to land on the surface of Mars, the *Viking 1* and *2* landers. Since it's more likely that life exists on Mars than the Moon, special precautions were taken to make sure that no terrestrial bacteria could contaminate Mars or its indigenous life forms. Before they were launched, the two *Viking* lander spacecraft were completely sterilized, and baked to a high temperature to kill off any terrestrial bacteria.

The idea of a sample return from Mars is scientifically very exciting, but the planetary protection aspects are also very serious. NASA is planning to construct a new facility that could carefully quarantine any samples robotic missions return from Mars until they can be completely studied and deemed not harmful to life on Earth. The second reason for care is to avoid the possibility of contaminating the samples with terrestrial bacteria that might be mistaken as Martian in origin.

Redefining the Habitable Zone

If we expand our definition of the habitable zone to include subsurface water, we can include Jupiter's moon Europa as a possible place for life to exist. In fact, after Mars, Europa is probably the second most likely place to find nonterrestrial life forms. Remember that Europa is about the size of Earth's Moon, and has no atmosphere. But beneath Europa's icy surface could lie an ocean of liquid water larger than all the oceans on Earth combined! Tidal heating from resonance with two other moons, as well as Jupiter's gravity, provides enough energy to keep water liquid, and might provide energy to cause volcanic activity at the bottom of the ocean. Since mid-ocean ridges are a prime location for terrestrial life on Earth, scientists are excited about the possibility of life far below the surface of Europa, although such life would likely still rely on materials produced at the surface in order to survive.

If Europa does have a global subsurface ocean, then the planetary protection scenario is even more important than for rocky planets like Mars. If a probe were to crash into Europa and accidentally penetrate to the ocean layer, it could easily contaminate the whole ocean (and therefore the whole potential biosphere), rather than just the immediate surroundings.

FACTS

The *Galileo* spacecraft, which has been in orbit around Jupiter since 1995, will nosedive into Jupiter and burn up some time in the next few years. To prevent *Galileo* from crash landing on Europa by mistake, scientists are sending *Galileo* deliberately into Jupiter rather than just letting its orbit decay naturally.

The possibility of life on Europa has caused astronomers to expand their definitions of the habitable zone. In addition to searching for planets in the Sun's own habitable zone, they are considering the possibility of life on satellites of planets which are either in or outside the habitable zone. Consider, for example, a huge Jupiter-sized planet orbiting a faraway star system at about the same distance Earth is from our Sun. The Jupiter-sized planet would be far too large for life to exist on it, but what about an Earth-sized satellite of that huge planet? That moon could be an excellent candidate for life.

SETI: The Search for Extraterrestrial Intelligence

If you combine the criteria listed for inhabitable planets, and expand them to take into account large moons of giant planets, then you have a good starting point for a search for life. Of course, it's hard enough to search for life on Mars and Europa, places in our own solar system. If we want to find life on planets in other solar systems, orbiting other stars, we need to use different methods. These other stars, for instance, are very far from our own solar system. Even the closest star system to us, the Alpha Centauri system, is about 4.3 light-years away from us. Traveling there at the speed of light, a feat currently far beyond our capabilities, would still take over four years!

Why Theory Is Important

Trying to find microscopic life forms on planets orbiting a far-off star is close to impossible at this point. We do not have sufficiently sensitive detectors to even observe extrasolar planets directly, let alone search for signs of microscopic life there! We are just able to find other solar systems and detect extrasolar planets, and we can theorize about which of them are more likely to contain life due to their sizes and locations in the habitable zone—but we cannot find evidence of microscopic, or even macroscopic, nonintelligent life.

FACTS

The movie *Contact* with Jodie Foster was based quite closely on some of the radio telescope SETI (Search for Extraterrestrial Intelligence) searches done at the SETI Institute, *www.seti.org*. Of course, no signals have actually been found from an extraterrestrial source yet!

Short of having extraterrestrials land on Earth and proclaim their existence, the best we can do currently in the search for non-Earth life is to search for non-Earth intelligent life. That's what SETI (Search for Extraterrestrial Intelligence) is all about. Even unintentionally, a technological civilization will send a variety of signals out into space, some of which can travel away from the planet in all directions and eventually be received by listeners on another world. In fact, we have been sending out such signals unintentionally for more than fifty years via high-frequency radio, television, and radar.

How SETI Was Founded

In 1960, terrestrial astronomers realized that if we were sending out signals to the stars, it was likely that other intelligent civilizations would also be sending out such signals, either intentionally or unintentionally. Frank Drake, a young radio astronomer, made the first search for signals coming from two nearby Sunlike stars by listening to them with a microwave radio antenna. He found no signals, but the prospect of listening for an extraterrestrial signal captured the excitement of scientists around the world, and they began listening programs of their own.

The wildly popular free screensaver SETI@home allows your personal computer to participate in the search for intelligent signals! Go to *http://setiathome.ssl.berkeley.edu* to download the screensaver, which helps process observing data from radio telescopes, and join over 3 million people worldwide who are helping out.

In designing a listening program, radio astronomers must select two things. First, they must pick which stars to observe. Should they keep their search as broad as possible and observe as many stars as they can? Or should they select a certain group of stars that could have a greater likelihood of success and observe only those, for a longer period of time?

When selecting a group of stars to observe, scientists use the same criteria described earlier. They want stars that are similar to our Sun—not too big and not too small, with a long enough lifetime to allow life to develop and flourish, old enough that intelligent life could already have arisen, and with a sufficient amount of heavier elements.

Second, a radio astronomer must select a certain wavelength at which to observe. Some surveys are able to scan through a huge number of wavelengths in order to catch signals at any possible frequency. Other surveys select one, or a few, fundamental frequencies that the scientists involved believe would be universal. One such frequency is the one used by Frank Drake in that first survey, and comes from the vibrations of the neutral hydrogen element. Since hydrogen is the most common element in the universe, astronomers believe that this wavelength might be used in an attempt to communicate.

SETI was originally funded by NASA, but funding was cut in the early 1990s. Since then, SETI has been conducted by private universities and research groups, including the nonprofit SETI Institute in Mountain View, California. The SETI Institute is currently conducting a multichannel radio search of 1,000 of the closest Sunlike stars, using the largest radio telescopes currently available here on Earth, including the immense Arecibo telescope in Puerto Rico. The SETI Institute is also working on building an array of small consumer satellite dishes that will work together as a giant radio telescope to look for signals (and perform conventional radio astronomy as well!).

Our Efforts to Contact Others

In addition to listening passively for signals from outer space, scientists have also sent out a few targeted messages intended especially

FIGURE 18-3:
This binary image was sent out by the Arecibo observatory in 1974

(refer to page 280 for more information)

Courtesy of NASA/Jet Propulsion Laboratory (JPL), Caltech

for intelligent extraterrestrial recipients. One such message was sent out in 1974 from the Arecibo observatory. This signal could be turned into a simple picture that described our solar system, the chemical compounds important to life, the structure of DNA, and the appearance of a human being. This signal was sent out in the direction of globular star cluster M13, which is 21,000 light-years away! Clearly, if we ever get an answer, it will be a long time in coming! The Pioneer 10 and 11 spacecraft both carried a simple pictorial plaque to show any intelligent life intercepting the spacecraft at some point in the distant future where Pioneer came from and who launched it.

A more sophisticated message to the stars was sent as part of the *Voyager 1* and *2* spacecraft. Each of these spacecraft carried a gold-plated 12 inch phonograph record, along with a needle and symbolic instructions for how to construct a device to play it. The record is engraved with pictures that show the structure of our solar system. Encoded on the record are 115 images, including mathematical definitions, pictures of the planets in our solar system, anatomy of humans, a variety of pictures of humans and other species on Earth, and natural and built scenes from around the world. The record also contains an audio portion consisting of a variety of natural sounds, music from around the world, and spoken greetings in fifty-five languages. The two *Voyager* spacecraft may make a close approach to another planetary system in about 40,000 years. The noted astronomer Carl Sagan (1934–1996) called the *Voyager* record a "message in a bottle."

FIGURE 18-4:
The *Pioneer 10* and *11* plaque

(refer to page 280 for more information)

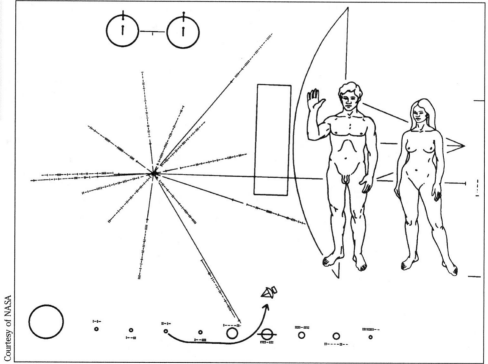

Courtesy of NASA

Thoughts to the Future

One thing that is quite unlikely, even though it's common in science fiction, is the effortless travel between stars over vast distances. Scientists still have not uncovered a way to travel faster than the speed of light, and such travel is impossible given what we know about the laws of physics. Warp speed is still an impossibility to us. And even traveling at the speed of light, the nearest star system would still be four years away!

Has Earth been visited by aliens?
Despite popular beliefs about alien abductions and UFO sightings, there is no reliable scientific evidence that extraterrestrials of any kind have ever visited Earth.

Other tricks used in science fiction, such as traveling through wormholes or along cosmic strings, seem equally improbable when you actually examine the physics. Unfortunately, the implication is that when we do finally try to travel outside our own solar system, it will likely be on ships that take many human generations to reach their destinations. Such a situation would demand either a robotic crew, or a crew of humans who will have to reproduce in order to survive until the ship arrives. This method of travel would still allow us to colonize the galaxy, although much more slowly than the Star Trek model. Communications would also take an extremely long time. It would be very difficult to build up a centrally controlled galactic empire using this method—instead, each colony would be basically autonomous. But until someone discovers a faster-than-light rocket drive, this model provides our only possibility for true space colonization.

CHAPTER 19

How It Will All End

Some say the world will end in fire, some say in ice.
—Robert Frost, "Fire and Ice"

This quote describes, fairly accurately, what will happen when the universe reaches its end. Consider the possible chain of events around the end of the habitability of the solar system: Although Earth is currently the most inviting and habitable planet in the solar system, it won't necessarily stay that way!

The Future Evolution of the Sun

As the Sun's evolution continues, it will slowly get hotter and hotter. Theories of the Sun's future vary, but we'll describe one such theory in detail here (with the caveat that the actual details may be somewhat different). When the Sun is about 5.6 billion years old (about 1.1 billion years from now), it will be about 10 percent brighter than it is now. This extra solar energy will result in a moist greenhouse effect, until Earth's atmosphere slowly dries out as water vapor is lost into space. This condition will probably signal the end of large surface life on Earth (like us!), but some marine life and simple land life forms may survive.

After about 9 billion years (about 4.5 billion years from now), the Sun will be about 40 percent brighter than it is now. This extra energy from the Sun will cause a runaway greenhouse effect on Earth, similar to what happened on the planet Venus. The oceans will heat up and evaporate; Earth will be left with a thick carbon dioxide atmosphere, almost no water, and a scorching hot surface. The end of all life on Earth will probably occur at this point.

The End of Hydrogen Fusion

Assuming that some future humans will be curious enough to stick around to see what happens next, the main event will occur when the Sun is about 10.9 billion years old and runs out of hydrogen fuel in its core. When the pressure from fusion decreases, the helium that has built up in the core will begin to collapse. This collapse will cause the core to heat up and become increasingly dense. The last bits of hydrogen left will continue fusion, but in a thin shell that surrounds the collapsing core of helium. At this point, the Sun will be about 1.5 times its current size, and over twice as bright. The Sun will then evolve slowly over about the next 700 million years or so. Its brightness will stay about the same, but its radius will slowly increase up to 2.3 times its current size, and its surface will start to cool.

At an age of 11.6 billion years, the Sun will become a subgiant star. At this point, the Sun will grow much faster, and at an age of about 12.1 billion years strong solar winds will carry almost 30 percent of the Sun's mass out into the solar system. A decreased solar mass means that the planets will have less of a gravitational bind to the Sun, and they will slowly move out in their orbits.

Our Sun's lifetime is half over, meaning we have only a few billion years before our planet becomes uninhabitable. First the Sun will gradually warm up, raising the surface temperature of Earth. Eventually it will expand into a red giant, enveloping the solar system's inner planets.

The Red Giant Phase

The Sun's size will increase until it reaches its maximum size in the red giant phase, at about 12.2 billion years. The Sun will expand to a radius of about 166 times its current size, and the huge bloated outer atmosphere of the Sun will pass the orbit of Mercury (which will be pulled into the Sun and destroyed). During this phase, the Sun will also be much brighter than it is currently (because it is so much larger in surface area)—about 2,300 times its current brightness. The overall surface temperature of the Sun will be lower, though, giving it its red color. The view from Earth during this phase, if anyone were around to see it, would be amazing—the red giant Sun would fill a huge portion of the sky.

When the Sun reaches its maximum size, the helium at its core will have reached a temperature sufficient for helium fusion to take place. The Sun will have a new temporary energy source, as helium fuses into carbon and oxygen. The Sun will stop expanding, and will begin to shrink in size over a period of about 1 million years.

At an age of about 12.2 billion years, the Sun will settle down for a short, stable period and burn helium. This stable period will be astronomically brief, however, and will last only about 110 million years.

At an age of about 12.3 billion years, the Sun will run out of helium in its core, and carbon and oxygen that are left will begin to collapse quickly (similar to what happened at the end of the hydrogen-burning phase). Another period of rapid growth will begin, but this time it will be much faster and take only 20 million years. However, the Sun's initial mass was too small for the core to get hot enough for another stage of fusion.

The Sun will enter a second red giant phase, but this time, the Sun's final size will be bigger (180 times its current radius), and it will get even brighter (3,000 times its current brightness). It again will lose much of its outer atmosphere as a strong solar wind develops, and eventually will lose almost half its original mass. Venus and Earth will then move outward just enough to avoid being swallowed up by the Sun—maybe.

Planetary Nebula

FIGURE 19-1:
Planetary nebula NGC-3132

(refer to page 280 for more information)

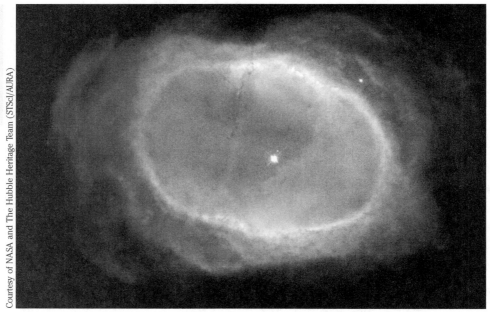

Courtesy of NASA and The Hubble Heritage Team (STScI/AURA)

Since the Sun won't be big enough for the next stage of fusion to ignite, the second red giant phase will signal the beginning of the end for our star. At an age of about 12.365 billion years, the Sun's interior will become unstable, as the remaining shell burning helium on top of the carbon and oxygen core begins to fluctuate in nuclear reaction rate, due mostly to changes in temperature. This stage will cause violent pulses to take place every 100,000 years. The last pulse will blow off the last of the Sun's atmosphere, leaving a hot inner core made up mostly of carbon and oxygen. The Sun will thereafter be no more.

The core will be about the size of Earth, and its temperature will start out very hot but will decrease rapidly as heat is lost to space, and there is no longer an energy source. The brightness of the core will also decrease rapidly as the Sun cools. It will eject ultraviolet light during this phase, and this action will light up the previously ejected envelope that used to be the Sun's atmosphere. This material will appear as a planetary nebula to life forms watching safely from other solar systems. The cloud of gas will last for only about 10,000 years, and will slowly dissipate into space. During this phase, however, the Sun will briefly appear as a very beautiful celestial object, before it loses all its heat and light. The leftover core of the Sun will form a white dwarf star, which will fade away until it is cold and dark, and virtually undetectable.

Effects on the Solar System

How will the Sun's changes affect the planets in the solar system? What will the fate be for life in the solar system? We know the fate of poor Mercury—it will be engulfed by the Sun during its first red giant phase. And Venus was never hospitable to begin with, so increased heat from the Sun will only make matters worse.

Earth and Mars

About a billion years from now, as the Sun's brightness increases, life on Earth will start to get unpleasant. The atmosphere will begin to contain more and more carbon dioxide, and plants and animals will have a difficult time surviving under these conditions. Two or three billion years from now, Earth will be a Venuslike, uninhabitable planet. But what's bad for Earth could be good for Mars.

Today, Mars is too cold and its atmosphere too thin to be hospitable for life. Although Mars may have once gone through a warmer, wetter period when water flowed on its surface, as it got colder, this water turned into ice and was trapped beneath the planet's surface or in the polar caps. As the Sun gets brighter and brighter, however, it could warm the surface of Mars enough to melt the polar caps and subsurface ice. This water could cover the surface, and the water vapor would begin to form an Earthlike atmosphere. Mars would develop only a transient hospitable phase, however. It might be a good refuge from a too-hot Earth, but as the Sun gets even hotter, eventually even Mars won't be a good place for us to survive.

Gas Giants

The gas giant planets of the outer solar system have very high gravity and no solid surfaces, and wouldn't be appropriate places for survival. But places like Jupiter's moon Europa could be a possibility. As the Sun heats up even more the icy surface of Europa could begin to melt. This process could produce an atmosphere, but Europa would be a tiny water world the size of our Moon—a possibility, but perhaps not too appealing for air-breathing humans.

Saturn

The Saturn system would be the next choice. Changes would already have taken place there. The rings of Saturn might slowly disappear, and might be completely gone after 100 million years if they are not sufficiently resupplied. Saturn's moon Titan could be a likely location for life to survive, but only temporarily. Titan is currently very cold, and is covered in a thick atmosphere. We know that organic chemical reactions are currently taking place in Titan's atmosphere; increased amounts of light from the Sun should help increase the rates of reactions taking place there. So even if Earth life doesn't find a safe haven on Titan, it's possible that primitive life could develop independently there as the Sun heats it up.

What's Left?

Even the Jupiter and Saturn systems will eventually become inhospitable, however, as the Sun heats up even more. Changes will also take place out at Uranus and Neptune, but there's little hope for finding a home for refugees there. Uranus has no promising large satellites, so it's unlikely this system could help us out. And only a billion years or so from now, Neptune's large moon Triton, which is slowly getting closer to the planet, will be so near that it will be ripped apart by Neptune's gravity. A beautiful ring system will likely be created surrounding Neptune, but it won't be any place to call home. Pluto, a small frozen iceball, isn't very appealing either.

After Mars is no longer habitable, the best place for life to survive could be the moons of the outer solar system.

Perhaps the best idea is to send out immense starships in all directions, like seeds blown to the wind, and hope that at least a few of them end up in a favorable location to take root and grow a new civilization. These solar system refugees may eventually be able to create new lives on a faraway world, and hopefully live there in peace for a few billion years before that star in turn reaches the end of its lifetime and they are forced to search for a new home yet again.

Destiny for the Universe

Even if humanity manages to escape the end of the habitability of our solar system, we can't possibly escape the end of the universe. We believe that the universe came into existence in a giant explosion nicknamed the Big Bang, but what will happen at the end of the universe's existence? What lies in store for humanity billions of years from now?

The universe is expanding, and the end result of this expansion depends on how much mass the universe has (just like the fate of a star depends on how massive it was initially). This mass limit is expressed in a critical density—if the density of the universe is less than a certain amount, then the universe's expansion will win out over gravity, and the universe will expand forever. This situation is called an *open universe.*

If the density of the universe happens to be exactly equal to the critical density, then the universe is called a *flat universe.* In this case, the velocity of expansion will very slowly approach zero, but never actually reach it. The expansion rate of the universe will slow down over time, but never actually stop. A flat universe will also expand forever, although not as quickly as an open universe.

ESSENTIALS   Robert Frost wrote, "Some say the world will end in fire, some say in ice." A closed universe is an end of the world in fire: the Big Crunch. The flat or open universe is the ice case—the universe will expand indefinitely, and in doing so will become colder and colder. We can call this option the Big Chill.

The third case is perhaps the most interesting. If the density of the universe is greater than the critical density, then expansion will eventually cease, and the universe will begin contracting again at some point. The resulting situation is called a *closed universe*, and in this case the universe will continue contracting until it reaches a point of maximum density again. A closed universe is finite both in space and time—it has a finite maximum size, and a finite maximum age.

The Closed Universe

Let's begin with the closed universe case. Remember that this is a universe whose density is over the critical density—here, we'll take this number to be about twice the critical density. Current estimates say the universe ranges from 10 billion to 20 billion years old. For the value of density we're considering, the universe will reach its maximum size approximately 50 billion years after the Big Bang (about 35 billion years from now). Expansion will cease at this point and, on average, galaxies will be about twice as far away from each other as they are now. After this point of maximum expansion is reached, the universe will begin collapsing.

Contraction Begins

At a point in time about 85 billion years after the Big Bang, this contraction will be well established. The distances between galaxies will have shrunk back to their current values, and the temperature of the cosmic microwave background radiation will have risen back up to its current value. This contraction will continue for the next 70 billion years or so, eventually speeding up as the universe increases in density and gravitational forces are increased. Densities and temperatures will then start to increase at a much faster rate.

FACTS

If there are any astronomers around studying the sky during contraction of the universe, they will be measuring blueshifts instead of redshifts, because all the objects in the galaxy will appear to be coming toward us rather than rushing away from us.

FIGURE 19-2:
Two merging galaxies

(refer to page 280 for more information)

Courtesy of NASA and The Hubble Heritage Team (STScl/AURA)

Galaxies Merge

When the background temperature reaches 50 Kelvin, the galaxies will all merge together into one giant megagalaxy since the average distance between galaxies will have shrunk to the size of a galaxy itself. This megagalaxy will contain all the stars in the universe. Once the background temperature of the universe rises much higher, the interstellar medium (leftover gas and dust located between stars) will be heated up enough to radiate energy on its own in the infrared. When the interstellar medium reaches a temperature hotter than the surface of a star, stars will no longer be able to radiate heat into space because they will actually be cooler than space! The heat and energy generated by stars will just be stuck inside, making their interiors hotter and hotter, and causing the pressure inside them to rise. Once the pressure rises too high to be contained, the star will explode.

The Big Crunch

The destruction of individual stars will be another step along the way toward making the universe homogenous. First individual galaxies will merge together, then individual stars will be destroyed and contribute their material to the ever-growing gas cloud. As the universe contracts even more, the Big Bang is actually run in reverse. This condition is called the Big Crunch.

Astronomers are still uncertain what exactly will happen at the end of the Big Crunch, though. Will the universe collapse down even further, forming a gigantic black hole? Or will the universe bounce back and start expanding again in another Big Bang, forming an entirely new universe?

The Open Universe

The other possibility for the end of the universe is infinite expansion. If the universe is open or flat, expansion will continue, and star formation and other processes that are taking place today in the universe will also continue. This universe will last much longer than the closed universe case. However, there is a finite amount of fuel in the universe; as time goes on, more and more of that fuel will be used up, and then star formation will cease.

The End of Starlight

In the open universe model, about 1 trillion years after the Big Bang, stars will die. Star formation will have already ceased because all the gas in the universe will be gone. Galaxies will be made up only of leftover remnants such as white dwarfs, neutron stars, and black holes. No more stars will shine in the sky.

Particles Decay

After many trillions of years, white dwarfs and neutron stars (the only two remaining "normal" types of objects left) will no longer be stable because the very particles making them up will have decayed! White dwarfs and neutron stars will then disintegrate into clouds of electrons, positrons, and photons. After this stage, the universe will again be stable for a long period of time, and will be made up of an expanding cloud of particles such as electrons, positrons, neutrinos, and photons, as well as stellar-mass black holes and supermassive black holes.

Black Holes Evaporate

Although in the classical view of physics black holes absorb particles but can't emit them, when you look at them under quantum mechanics they actually seem to emit particles if a particle-antiparticle pair comes into existence very close to a black hole. In this case, if one of the pair is sucked into the black hole, the other particle seems to be emitted by the black hole as it escapes. This emission results in a very slow energy loss from the black hole that eventually leads to it reaching a mass of zero and disappearing.

QUESTIONS?

How big is supermassive?
The term sounds exaggerated, but think of it in relative terms. Stellar-mass black holes would have come from the collapse of massive stars. Supermassive black holes were once the centers of entire galaxies.

Finally, the last remnant of structure in the universe, the supermassive black holes, also evaporate through the same process described for stellar-mass black holes. At this point, the entire universe is a still-expanding cloud of electrons, positrons, photons, and neutrinos. As far as we know, none of these particles will ever decay, so this final configuration of the universe is stable.

Particle emissions from black holes, and the energy loss that results, take a very long time—about 10^{65} years for a black hole whose initial mass was that of our Sun.

The universe continues expanding, however, and the temperature becomes lower and lower. This condition is what is sometimes called the Big Chill. This situation, the slow dissolution of the ordered universe into a huge, homogeneous sea of particles not doing anything at all, can also be called the Big Bore. Remember that the flat universe model is very similar to the open universe model, and will have just about the same final result, except that the expansion of the flat universe will get slower and slower over time, but will never actually stop.

Our Fate

So, which of these scenarios is ahead in our future? Will our universe eventually contract a primordial fireball in the Big Crunch, or will it slowly expand and spread out into a sea of nothingness?

The answer to these questions depends on the density of the universe, an area of current research. When simply adding up all the visible mass in the universe, scientists arrive at a value far below the critical density described earlier. This estimate would put us in an open universe, expanding forever into a lower and lower density cloud of primitive particles. But is this scenario really the case? The main problem is that astronomers believe that much of the mass of the universe could be made up of so-called dark matter, which is basically undetectable because it gives off no radiation.

Dark Matter

Why do astronomers believe there could be dark matter lurking out there, unseen in the skies? How much of it could be out there? Based on observations of the motions of galaxies and the expansion rate of the universe, astronomers estimate that as much as 90 percent of the mass in the universe is technically dark matter, and does not radiate at all.

What Is Dark Matter?

Astronomers have proposed a variety of types of material that could be classified as dark matter. It could consist of very faint stars, or brown dwarfs (almost-stars that weren't quite large enough to ignite), or even burnt-out white dwarf stars that have radiated the last of their heat into space and are now indistinguishable from the background temperature of their galaxy.

Dark matter could mostly consist of black holes or cold nonradiating clouds of gas or dust. Some astronomers have proposed objects called MACHOs (Massive Compact Halo Objects) surrounding galaxies and galactic clusters, which could be made up of brown dwarfs and other similar objects.

Another possibility is that dark matter is made up of strange, exotic particles that we haven't yet been able to observe. Such particles have been in the realm of theoretical physics so far, and they have not been confirmed to exist even in laboratory experiments.

How Can We See It?

Since dark matter gives off no radiation of its own, it is impossible to observe directly. But we can observe it indirectly—we can look for the effects that it has on other objects that we can observe, similar to the same kinds of detective techniques used to find black holes (in fact, black holes are a kind of dark matter). For instance, as astronomers observe the motions of stars and gas clouds in spiral galaxies, they can see that these objects move faster than what would be predicted based solely on the expected mass of the galaxy. This result means that there is more mass in the galaxy than we can observe, thereby increasing the galaxy's gravitational pull. This mass could be located in a spherical halo surrounding the galaxy, and could contain 90 percent of the galaxy's total mass.

Even when current estimates of dark matter are all taken into account, the density of the universe that astronomers come up with is still less than the critical density. When astronomers add up all the visible matter, plus all the dark matter they theorize exists, they come up with only about 20 percent of the critical density. This tally means that the universe doesn't have enough matter to keep from expanding forever, and that we are indeed in an open universe headed for the Big Chill.

Theory Versus Reality

Although most theoretical astronomers prefer the density of their model universes to be exactly the critical density, there is little evidence that the real universe follows their models. Substantial amounts of missing mass, in the form of dark matter, would have to be found for the universe to reach this critical density, and even more would be necessary for the universe to be closed. It seems likely that our universe is an open one and that it will expand forever, breaking down further and further into smaller and smaller particles that eventually spread out into a uniform, low-density cloud.

This view provides an interesting contrast to the end of our planet. Earth likely will end in fire, either when it is consumed by our Sun during one of its red giant phases, or when its surface gets fried and scorched if it manages to survive this phase. Ultimately, however, our universe will end in ice—the cold, unmoving ice of an ever-expanding, noninteracting sea of simple particles doing next to nothing at all. Which would you prefer?

Appendices

Constellation List

As looking up on a clear night becomes more of a habit for you, you may find yourself interested in checking out other skies, or your own during other seasons. Use the following information to supplement your star map or planisphere and get the most from your observing experience.

CONSTELLATION LIST

NUMBER	CONSTELLATION	ENGLISH NAME	HEMISPHERE	ALPHA STAR
1	Andromeda	Andromeda	NH	Alpheratz
2	Antlia	Air Pump	SH	
3	Apus	Bird of Paradise	SH	
4	Aquarius	Water Carrier	SH	Sadalmelik
5	Aquila	Eagle	NH/SH	Altair
6	Ara	Altar	SH	
7	Aries	Ram	NH	Hamal
8	Auriga	Charioteer	NH	Capella
9	Bootes	Herdsman	NH	Arcturus
10	Caelum	Chisel	SH	
11	Camelopardalis	Giraffe	NH	
12	Cancer	Crab	NH	Acubens
13	Canes Venatici	Hunting Dogs	NH	Cor Caroli
14	Canis Major	Big Dog	SH	Sirius
15	Canis Minor	Little Dog	NH	Procyon
16	Capricornus	Goat	SH	Algedi
17	Carina	Keel	SH	Canopus
18	Cassiopeia	Cassiopeia	NH	Schedar
19	Centaurus	Centaur	SH	Rigil Kentaurus
20	Cepheus	Cepheus	SH	Alderamin
21	Cetus	Whale	SH	Menkar
22	Chamaleon	Chameleon	SH	
23	Circinus	Compasses	SH	
24	Columba	Dove	SH	Phact
25	Coma Berenices	Berenice's Hair	NH	Diadem
26	Corona Australis	Southern Crown	SH	

CONSTELLATION LIST (continued)

NUMBER	CONSTELLATION	ENGLISH NAME	HEMISPHERE	ALPHA STAR
27	Corona Borealis	Northern Crown	NH	Alphecca
28	Corvus	Crow	SH	Alchiba
29	Crater	Cup	SH	Alkes
30	Crux	Southern Cross	SH	Acrux
31	Cygnus	Swan	NH	Deneb
32	Delphinus	Dolphin	NH	Sualocin
33	Dorado	Goldfish	SH	
34	Draco	Dragon	NH	Thuban
35	Equuleus	Little Horse	NH	Kitalpha
36	Eridanus	River	SH	Achernar
37	Fornax	Furnace	SH	
38	Gemini	Twins	NH	Castor
39	Grus	Crane	SH	Al Na'ir
40	Hercules	Hercules	NH	Rasalgethi
41	Horologium	Clock	SH	
42	Hydra (female)	Sea Serpent	SH	Alphard
43	Hydrus (male)	Water Serpent	SH	
44	Indus	Indian	SH	
45	Lacerta	Lizard	NH	
46	Leo	Lion	NH	Regulus
47	Leo Minor	Smaller Lion	NH	
48	Lepus	Hare	SH	Arneb
49	Libra	Balance	SH	Zubenelgenubi
50	Lupus	Wolf	SH	Men
51	Lynx	Lynx	NH	
52	Lyra	Lyre	NH	Vega
53	Mensa	Table	SH	
54	Microscopium	Microscope	SH	
55	Monoceros	Unicorn	SH	
56	Musca	Fly	SH	
57	Norma	Square	SH	
58	Octans	Octant	SH	

CONSTELLATION LIST (continued)

NUMBER	CONSTELLATION	ENGLISH NAME	HEMISPHERE	ALPHA STAR
59	Ophiucus	Serpent Holder	NH/SH	Rasalhague
60	Orion	Orion	NH/SH	Betelgeuse
61	Pavo	Peacock	SH	Peacock
62	Pegasus	Winged Horse	NH	Markab
63	Perseus	Perseus	NH	Mirfak
64	Phoenix	Phoenix	SH	Ankaa
65	Pictor	Easel	SH	
66	Pisces	Fishes	NH	Alrischa
67	Pisces Austrinus	Southern Fish	SH	Fomalhaut
68	Puppis	Stern	SH	
69	Pyxis	Compass	SH	
70	Reticulum	Reticle	SH	
71	Sagitta	Arrow	NH	
72	Sagittarius	Archer	SH	Rukbat
73	Scorpius	Scorpion	SH	Antares
74	Sculptor	Sculptor	SH	
75	Scutum	Shield	SH	
76	Serpens	Serpent	NH/SH	Unuck al Hai
77	Sextans	Sextant	SH	
78	Taurus	Bull	NH	Aldebaran
79	Telescopium	Telescope	SH	
80	Triangulum	Triangle	NH	Ras al Mothallah
81	Triangulum Australe	Southern Triangle	SH	Atria
82	Tucana	Toucan	SH	
83	Ursa Major	Great Bear	NH	Dubhe
84	Ursa Minor	Little Bear	NH	Polaris
85	Vela	Sails	SH	
86	Virgo	Virgin	NH/SH	Spica
87	Volans	Flying Fish	SH	
88	Vulpecula	Fox	NH	

CONSTELLATIONS BY MONTH

MONTH	CONSTELLATIONS
January	Caelum, Dorado, Mensa, Orion, Reticulum, Taurus
February	Auriga, Camelopardalis, Canis Major, Columba, Gemini, Lepus, Monoceros, Pictor
March	Cancer, Canis Minor, Carina, Lynx, Puppis, Pyxis, Vela, Volans
April	Antlia, Chamaeleon, Crater, Hydra, Leo, Leo Minor, Sextans, Ursa Major
May	Canes Venatici, Centaurus, Coma Berenices, Corvus, Crux, Musca, Virgo
June	Boötes, Circinus, Libra, Lupus, Ursa Minor
July	Apus, Ara, Corona Borealis, Draco, Hercules, Norma, Ophiuchus, Scorpius, Serpens, Triangulum Australe
August	Corona Australis, Lyra, Sagittarius, Scutum, Telescopium
September	Aquila, Capricornus, Cygnus, Delphinus, Equuleus, Indus, Microscopium, Pavo, Sagitta, Vulpecula
October	Aquarius, Cepheus, Grus, Lacerta, Octans, Pegasus, Piscis Austrinus
November	Andromeda, Cassiopeia, Phoenix, Pisces, Sculptor, Tucana
December	Aries, Cetus, Eridanus, Fornax, Horologium, Hydrus, Perseus, Triangulum

Planetary Facts

C hapters 6 and 7 cover the truly fascinating aspects of the major bodies in our solar system, but sometimes you just want the facts. The following table can help you understand how the location, size, and mass of the planets compare.

PLANETARY FACTS

PLANET	ORBIT	DIAMETER	MASS
Mercury	57,910,000 km (0.38 AU) from Sun	4,880 km	3.30×10^{23} kg
Venus	108,200,000 km (0.72 AU) from Sun	12,103.6 km	4.869×10^{24} kg
Earth	149,600,000 km (1.00 AU) from Sun	12,756.3 km	5.972×10^{24} kg
Mars	227,940,000 km (1.52 AU) from Sun	6,794 km	6.4219×10^{23} kg
Jupiter	778,330,000 km (5.20 AU) from Sun	142,984 km (equatorial)	1.900×10^{27} kg
Saturn	1,429,400,000 km (9.54 AU) from Sun	120,536 km (equatorial)	5.68×10^{26} kg
Uranus	2,870,990,000 km (19.218 AU) from Sun	51,118 km (equatorial)	8.683×10^{25} kg
Neptune	4,504,000,000 km (30.06 AU) from Sun	49,532 km (equatorial)	1.0247×10^{26} kg
Pluto	5,913,520,000 km (39.5 AU) from Sun (average)	2,274 km	1.27×10^{22} kg

AU: Astronomical Unit is the unit of measure equal to Earth's mean, or average, distance from the Sun. Hence, Earth is 1 AU from the Sun.

Appendix C
Messier's Catalogue

Charles Messier was the first to create a catalog of objects that were not comets. Although other catalogs have been created since, Messier objects are among the most popular (and visible) deep sky objects you can observe.

NUMBER	RIGHT ASCENSION	DECLINATION	VISUAL MAGNITUDE	COMMON NAME/DESCRIPTION
		MESSIER'S CATALOGUE		
M1	5h 20m 02s	+21d 45' 17"	8.4	Crab Nebula
M2	21h 21m 08s	−1d 47' 00"	6.5	Globular Cluster, Aquarius
M3	13h 31m 25s	+29d 32' 57"	6.2	Glob Cluster, Canes Venatici
M4	16h 09m 08s	−25d 55' 40"	5.6	Glob Cluster, Scorpius
M5	15h 06m 36s	+2d 57' 16"	5.6	Glob Cluster, Serpens
M6	17h 24m 42s	−32d 10' 34"	5.3	Open Cluster, Scorpius
M7	17h 38m 02s	−34d 40' 34"	4.1	Open Cluster, Scorpius
M8	17h 49m 58s	−24d 21' 10"	6.0	Lagoon Nebula, Sagittarius
M9	17h 05m 22s	−18d 13' 26"	7.7	Glob Cluster, Ophiuchus
M10	16h 44m 48s	−3d 42' 18"	6.6	Glob Cluster, Ophiuchus
M11	18h 30m 23s	−6d 31' 01"	6.3	Open Cluster, Scutum
M12	16h 34m 53s	−2d 30' 28"	6.7	Glob Cluster, Ophiuchus
M13	16h 33m 15s	+36d 54' 44"	5.8	Hercules Globular Cluster
M14	17h 25m 14s	−3d 05' 45"	7.6	Glob Cluster, Ophiuchus
M15	21h 18m 41s	+10d 40' 03"	6.2	Glob Cluster, Pegasus
M16	18h 05m 00s	−13d 51' 44"	6.4	Open Cluster, Serpens
M17	18h 07m 03s	−16d 14' 44"	7.0	Omega Nebula, Sagittarius
M18	18h 06m 16s	−17d 13' 14"	7.5	Open Cluster, Sagittarius
M19	16h 48m 07s	−25d 54' 46"	6.8	Glob Cluster, Ophiuchus
M20	17h 48m 16s	−22d 59' 10"	9.0	Trifid Nebula
M21	17h 50m 07s	−22d 31' 25"	6.5	Open Cluster, Sagittarius
M22	18h 21m 55s	−24d 06' 11"	5.1	Glob Cluster, Sagittarius
M23	17h 42m 51s	−18d 45' 55"	6.9	Open Cluster, Sagittarius

MESSIER'S CATALOGUE (continued)

NUMBER	RIGHT ASCENSION	DECLINATION	VISUAL MAGNITUDE	COMMON NAME/DESCRIPTION
M24	18h 01m 44s	−18d 26′ 00″	4.6	Sagittarius Star Cloud
M25	18h 17m 40s	−19d 05′ 00″	6.5	Open Cluster, Sagittarius
M26	18h 32m 22s	−9d 38′ 14″	8.0	Open Cluster, Scutum
M27	19h 49m 27s	+22d 04′ 00″	7.4	Dumbbell Nebula
M28	18h 09m 58s	−24d 57′ 11″	6.8	Glob Cluster, Sagittarius
M29	20h 15m 38s	+37d 11′ 57″	7.1	Open Cluster, Cygnus
M30	21h 27m 05s	−24d 19′ 04″	7.2	Glob Cluster, Capricornus
M31	0h 29m 46s	+39d 09′ 32″	3.4	Andromeda Galaxy
M32	0h 29m 50s	+38d 45′ 34″	8.1	Andromeda Satellie
M33	1h 40m 37s	+29d 32′ 25″	5.7	Triangulum Galaxy
M34	2h 27m 27s	+41d 39′ 32″	5.5	Open Cluster, Perseus
M35	5h 54m 41s	+24d 33′ 30″	5.3	Open Cluster, Gemini
M36	5h 20m 47s	+34d 08′ 06″	6.3	Open Cluster, Auriga
M37	5h 37m 01s	+32d 11′ 51″	6.2	Open Cluster, Auriga
M38	5h 12m 41s	+36d 11′ 51″	7.4	Open Cluster, Auriga
M39	21h 23m 49s	+47d 25′ 00″	5.2	Open Cluster, Cygnus
M40	12h 11m 02s	+59d 23′ 50″	8.4	Winnecke 4
M41	6h 35m 53s	−20d 33′ 00″	4.6	Open Cluster, Canis Major
M42	5h 23m 59s	−5d 34′ 06″	4.0	Orion Nebula
M43	5h 24m 12s	−5d 26′ 37″	9.0	De Mairan's Nebula
M44	8h 07m 22s	+20d 31′ 38″	3.7	Beehive Cluster, Praesepe
M45	3h 33m 48s	+23d 22′ 41″	1.6	Pleiades
M46	7h 31m 11s	−14d 19′ 07″	6.0	Open Cluster, Puppis
M47	7h 44m 16s	−14d 50′ 08″	5.2	Open Cluster, Puppis
M48	8h 02m 24s	−1d 16′ 42″	5.5	Open Cluster, Hydra
M49	12h 17m 48s	+9d 16′ 09″	8.4	Elliptical Galaxy, Virgo
M50	6h 51m 50s	−7d 57′ 42″	6.3	Open Cluster, Monoceros
M51	13h 20m 23s	+48d 24′ 24″	8.4	Whirlpool Galaxy
M52	23h 14m 38s	+60d 22′ 12″	7.3	Open Cluster, Cassiopeia
M53	13h 02m 02s	+19d 22′ 44″	7.6	Glob Cluster, Coma Berenices

MESSIER'S CATALOGUE (continued)

NUMBER	RIGHT ASCENSION	DECLINATION	VISUAL MAGNITUDE	COMMON NAME/DESCRIPTION
M54	18h 40m 52s	−30d 44' 01"	7.6	Glob Cluster, Sagittarius
M55	19h 26m 02s	−31d 26' 27"	6.3	Glob Cluster, Sagittarius
M56	19h 08m 00s	+29d 48' 14"	8.3	Glob Cluster, Lyra
M57	18h 45m 21s	+32d 46' 03"	8.8	Ring Nebula
M58	12h 26m 30s	+13d 02' 42"	9.7	Spiral Galaxy, Virgo
M59	12h 30m 47s	+12d 52' 36"	9.6	Elliptical Galaxy, Virgo
M60	12h 32m 28s	+12d 46' 02"	8.8	Elliptical Galaxy, Virgo
M61	12h 10m 44s	+5d 42' 05"	9.7	Spiral Galaxy, Virgo
M62	16h 47m 14s	−29d 45' 30"	6.5	Glob Cluster, Ophiuchus
M63	13h 04m 22s	+43d 12' 37"	8.6	Sunflower Galaxy
M64	12h 45m 51s	+22d 52' 31"	8.5	Blackeye Galaxy
M65	11h 07m 24s	+14d 16' 08"	9.3	Spiral Galaxy, Leo
M66	11h 08m 47s	+14d 12' 21"	8.9	Spiral Galaxy, Leo
M67	8h 36m 28s	+12d 36' 38"	6.1	Open Cluster, Cancer
M68	12h 27m 38s	−25d 30' 20"	7.8	Glob Cluster, Hydra
M69	18h 16m 47s	−32d 31' 45"	7.6	Glob Cluster, Sagittarius
M70	18h 28m 53s	−32d 31' 07"	7.9	Glob Cluster, Sagittarius
M71	19h 43m 57s	+18d 13' 00"	8.2	Glob Cluster, Sagitta
M72	20h 41m 23s	−13d 20' 51"	9.3	Glob Cluster, Aquarius
M73	20h 46m 52s	−13d 28' 40"	9.0	Asterism of 4 Stars, Aquarius
M74	1h 24m 57s	+14d 39' 35"	9.4	Spiral Galaxy, Pisces
M75	19h 53m 10s	−22d 32' 23"	8.5	Glob Cluster, Sagittarius
M76	1h 28m 43s	+50d 28' 48"	10.1	Little Dumbbell Nebula
M77	2h 31m 30s	−0d 57' 43"	8.9	Cetus A
M78	5h 35m 34s	−0d 01' 23"	8.3	Diffuse Nebula, Orion
M79	5h 15m 16s	−24d 42' 57"	7.7	Glob Cluster, Lepus
M80	16h 04m 00s	−22d 25' 13"	7.3	Glob Cluster, Scorpius
M81	9h 37m 51s	+70d 07' 24"	6.9	Bode's Galaxy
M82	9h 37m 57s	+70d 44' 27"	8.4	Cigar Galaxy
M83	13h 24m 33s	−28d 42' 27"	7.6	Southern Pinwheel

MESSIER'S CATALOGUE (continued)

NUMBER	RIGHT ASCENSION	DECLINATION	VISUAL MAGNITUDE	COMMON NAME/DESCRIPTION
M84	12h 14m 01s	+14d 07' 01"	9.1	Lenticular Galaxy, Virgo
M85	12h 14m 21s	+19d 24' 26"	9.1	Lent. Galaxy, Coma Berenices
M86	12h 15m 05s	+14d 09' 52"	8.9	Lenticular Galaxy, Virgo
M87	12h 19m 48s	+13d 38' 01"	8.6	Virgo A
M88	12h 21m 03s	+15d 37' 51"	9.6	Spiral Galaxy, Coma Berenices
M89	12h 24m 38s	+13d 46' 49"	9.8	Elliptical Galaxy, Virgo
M90	12h 25m 48s	+14d 22' 50"	9.5	Spiral Galaxy, Virgo
M91	12h 26m 28s	+14d 57' 06"	10.2	Spiral Galaxy, Coma Berenices
M92	17h 10m 32s	+43d 21' 59"	6.4	Globular Cluster, Hercules
M93	7h 35m 14s	−23d 19' 45"	6.0	Open Cluster, Puppis
M94	12h 40m 43s	+42d 18' 43"	8.2	Spiral Galaxy, Canes Venatici
M95	10h 32m 12s	+12d 50' 21"	9.7	Spiral Galaxy, Leo
M96	10h 35m 05s	+12d 58' 09"	9.2	Spiral Galaxy, Leo
M97	11h 01m 15s	+56d 13' 30"	9.9	Owl Nebula
M98	12h 03m 23s	+16d 08' 15"	10.1	Spiral Galaxy, Coma Berenices
M99	12h 07m 41s	+15d 37' 12"	9.9	Spiral Galaxy, Coma Berenices
M100	12h 11m 57s	+16d 59' 21"	9.3	Spiral Galaxy, Coma Berenices
M101	13h 43m 28s	+55d 24' 25"	7.9	Pinwheel Galaxy
M102	Possible duplication of M101			
M103	01h 33m 2s	+60d 42'	7.4	Open Cluster, Cassiopeia
M104	12h 28m 39s	−10d 24' 49"	8.0	Spiral Galaxy, Virgo
M105	10h 47m 8s	+12d 35'	9.3	Elliptical Galaxy, Leo
M106	12h 19m 0s	+47d 18'	8.4	Spiral Galaxy, Canes Venatici
M107	16h 32m 5s	−13d 03'	7.9	Glob Cluster, Ophiuchus
M108	11h 11m 5s	+55d 40'	10.0	Spiral Galaxy, Ursa Major
M109	11h 57m 6s	+53d 23'	9.8	Spiral Galaxy, Ursa Major
M110	00h 40m 4s	+41d 41'	8.5	Andromeda Satellie

Messier's Catalogue footnote: The faintest objects visible to the naked eye have a magnitude of around 6. The brighter something is, the lower its magnitude.

Web Sites for Amateur Astronomers

The following list of astronomy clubs and planetariums is composed of groups with permanent Web sites; although there are many other groups, a lack of a reliable Web site could make locating them quite frustrating. If you need more listings, check your local telephone book or the astronomy club locator on the Sky and Telescope Web site at *www.skypub.com*.

Astronomy Clubs
In the United States

ALABAMA
Birmingham Astronomical Society,
 www.bas-astro.com
Von Braun Astronomical Society, *www.vbas.org*

ARIZONA
Discovery Park Astronomical Society,
 www.discoverypark.com
East Valley Astronomy Club,
 www.eastvalleyastronomy.org
Saguaro Astronomy Club, *www.saguaroastro.org*
Tucson Amateur Astronomy Association,
 www.tucsonastronomy.org

ARKANSAS
CASE, *www.christian-astronomy.org*

CALIFORNIA
Antelope Valley Astronomy Club,
 www.avac.av.org
Astronomical Society of the Desert,
 www.astrorx.org
The Astronomy Connection,
 http://www.observers.org
Eastbay Astronomical Society,
 www.eastbayastro.org

FirstLight Astronomy Club,
 www.firstlightastro.com
Los Angeles Astronomical Society, *www.laas.org*
Mount Wilson Observatory Association,
 www.mwoa.org
Sacramento Valley Astronomical Society,
 www.skywatchers.org
San Diego Astronomy Association,
 www.sdaa.org
San Jose Astronomical Association,
 www.sjaa.net
Sidewalk Astronomers,
 www.sidewalkastronomers.com
Ventura County Astronomical Society,
 www.vcas.org

COLORADO
Aurora Astronomical Association,
 http://a_cubed.tripod.com
Colorado Springs Astronomical Society,
 www.rmss.org
International Association for Astronomical Studies,
 www.iaas.org

CONNECTICUT
Astronomical Society of New Haven,
 www.asnh.org
Westport Astronomical Society,
 www.was.visionnet.com

DELAWARE

Delaware Astronomical Society,
 www.cis.udel.edu/~case/das.html
Delmarva Stargazers, *www.delmarvastargazers.org*

DISTRICT OF COLUMBIA

National Capital Astronomers,
 http://capitalastronomers.org

FLORIDA

Alachua Astronomy Club,
 www.astro.ufl.edu/aac
Central Florida Astronomical Society,
 www.physics.ucf.edu/cfas
Escambia Amateur Astronomers' Association,
 www.eaaa.net
Southern Cross Astronomical Society,
 www.scas.org
Tallahassee Astronomical Society,
 www.stargazers.org

GEORGIA

Astronomy Club of Augusta,
 www.angelfire.com/ga/astronomyclubaugusta
Atlanta Astronomy Club,
 www.atlantaastronomy.org

HAWAII

Hawaiian Astronomical Society,
 www.hawastsoc.org
Meteor Group Hawaii, *www.meteor-group.com*

IDAHO

Boise Astronomical Society, *www.boiseastro.org*
Magic Valley Astronomical Society,
 www.mvas.net

ILLINOIS

Chicago Astronomical Society,
 www.chicagoastro.org
Fox Valley Astronomical Society, *www.fvastro.org*

Lake County Astronomical Society,
 www.lcas-astronomy.org
Peoria Astronomical Society,
 www.astronomical.org
St. Louis Astronomical Society,
 www.slasonline.org

INDIANA

Calumet Astronomical Society,
 www.casonline.org
Wabash Valley Astronomical Society,
 www.stargazing.net/wvas

IOWA

Cedar Amateur Astronomers,
 www.cedar-astronomers.org

KANSAS

Heartland Astronomical Research Team,
 www.skygazer.org/hart
Northeast Kansas Amateur Astronomers' League,
 www.kansas.net/~farpoint

KENTUCKY

Hilltopper Astronomy Club,
 http://hac.wku.edu
Louisville Astronomical Society,
 www.louisville-astro.org
Big South Fork Star Gazers,
 http://sfsg.8k.com

LOUISIANA

Pontchartrain Astronomy Society,
 www.nightskydesign.com/pas/index.html
Shreveport-Bossier Astronomical Society,
 www.lsus.edu/nonprofit/sbas

MAINE

Astronomical Society of Northern New England,
 www.asnne.org

Downeast Amateur Astronomers Club,
    www.deaa.2ya.com

MARYLAND

Baltimore Astronomical Society, *www.baltastro.org*
Cumberland Astronomy Club,
    http://antoine.fsu.umd.edu/phys/luzader/cac
Westminster Astronomical Society,
    www.westminsterastro.org

MASSACHUSETTS

Amateur Telescope Makers of Boston,
    www.atmob.org
Amherst Astronomy Association, *www.amastro.org*
Boston University Astronomical Society,
    www.bu.edu/astronomy/buas
Springfield Stars Club, *www.reflector.org*

MICHIGAN

Capital Area Astronomy Club,
    www.pa.msu.edu/abrams/astronomyclub
Grand Rapids Amateur Astronomical Association,
    www.graaa.org
Muskegon Astronomical Society,
    www.wmich-astro.org
Oakland Astronomy Club,
    www.surmount.com/oac
Shoreline Amateur Astronomical Association,
    www.macatawa.org/~saaa

MINNESOTA

Minnesota Astronomical Society, *www.mnastro.org*

MISSISSIPPI

Jackson Astronomical Association,
    http://jackson.astronomers.org

MISSOURI

Astronomical Society of Kansas City,
    www.askconline.org/index.htm

MONTANA

Mission Valley Astronomy Club,
    http://mvac.homestead.com/index.html

NEBRASKA

Omaha Astronomical Society,
    www.omahaastro.com
Prairie Astronomy Club,
    www.prairieastronomyclub.org

NEVADA

Las Vegas Astronomical Society,
    www.ccsn.nevada.edu/LVAS

NEW HAMPSHIRE

New Hampshire Astronomical Society,
    www.nhastro.com

NEW JERSEY

Amateur Astronomers Association of Princeton,
    www.princetonastronomy.org
Amateur Astronomers, Inc. *www.asterism.org*
New Jersey Astronomical Association,
    www.njaa.org
Rockland Astronomy Club,
    www.rocklandastronomy.com
Star Astronomy Club, *www.starastronomy.org*
Willingboro Astronomical Society,
    www.wasociety.org

NEW MEXICO

Alamogordo Astronomy Club,
    www.zianet.com/aacwp
Astronomical Society of Las Cruces,
    www.zianet.com/aslc
New Mexico Tech Astronomy Club,
    www.nmt.edu/~astro

NEW YORK

Amateur Astronomer's Association, *www.aaa.org*

Syracuse Astronomical Society,
 ✍ www.syracuse-astro.org

NORTH CAROLINA
Greensboro Astronomy Club,
 ✍ www.greensboro.com/astronomy
Raleigh Astronomy Club, ✍ http://rtpnet.org/~rac

NORTH DAKOTA
Northern Sky Astronomical Society,
 ✍ www.und.edu/org/nsas

OHIO
Cincinnati Astronomical Society, ✍ www.cinastro.org
Columbus Astronomical Society, ✍ www.the-cas.org
Mahoning Valley Astronomical Society,
 ✍ www.mvobservatory.com
Miami Valley Astronomical Society, ✍ www.mvas.org
Wilderness Center Astronomy Club,
 ✍ www.twcac.org/

OKLAHOMA
Astronomy Club of Tulsa, ✍ http://astrotulsa.com

OREGON
Eugene Astronomical Society,
 ✍ www.efn.org/~eugastso
Rose City Astronomers, ✍ www.rca-omsi.org/rca
Southern Oregon Skywatchers,
 ✍ www.oregonskywatchers.org

PENNSYLVANIA
Amateur Astronomers Association of Pittsburgh,
 ✍ http://trfn.clpgh.org/aaap
American Lunar Society,
 ✍ http://otterdad.dynip.com/als
Astronomical Society of Harrisburg,
 ✍ www.astrohbg.org
Central Pennsylvania Observers, ✍ www.cpoclub.org
Delaware Valley Amateur Astronomers,
 ✍ http://dvaa.org

Lehigh Valley Amateur Astronomical Society,
 ✍ www.lvaas.org
Oil Region Astronomical Society, ✍ www.oras.org

RHODE ISLAND
Skyscrapers, ✍ www:theskyscrapers.org

SOUTH CAROLINA
Midlands Astronomy Club,
 ✍ http//:astro.physics.sc.edu/Mac
Roper Mountain Astronomers, ✍ www.rmastro.com

SOUTH DAKOTA
Black Hills Astronomical Society,
 ✍ www.sdsmt.edu/space/BHAS.htm

TENNESSEE
Barnard Astronomical Society,
 ✍ www.chattanooga.net/bas
Barnard-Seyfert Astronomical Society,
 ✍ www.bsasnashville.com
Bays Mountain Astronomy Club,
 ✍ www.baysmountain.com
Memphis Astronomical Society,
 ✍ www.memphisastro.org

TEXAS
Austin Astronomical Society, ✍ www.austinastro.org
Houston Astronomical Society,
 ✍ www.astronomyhouston.org
North Houston Astronomy Club,
 ✍ http://astronomyclub.org
Olympus Mons Astronomical Society,
 ✍ www.olympusmons.org
Texas Astronomical Society, ✍ www.texasastro.org

UTAH
Ogden Astronomical Society,
 ✍ http://physics.weber.edu/oas/oas.html
Salt Lake Astronomical Society, ✍ http://slas.wsl

VERMONT

Springfield Telescope Makers,
  ✍ *www.stellafane.com*
Vermont Astronomical Society,
  ✍ *www.uvm.edu/~jrs/vas*

VIRGINIA

Back Bay Amateur Astronomers,
  ✍ *http://groups.hamptonroads.com/bbaa*
Northern Virginia Astronomy Club,
  ✍ *www.novac.com*
Society for Scientific Exploration,
  ✍ *www.scientificexploration.org*

WASHINGTON

Seattle Astronomical Society,
  ✍ *www.seattleastro.org*
Spokane Astronomical Society,
  ✍ *www.spokaneastronomical.org*
Tri-City Astronomy Club,
  ✍ *www.stargazing.net/TCAC*
Yakima Valley Astronomy Club,
  ✍ *www.perr.com/yvac.html*

WEST VIRGINIA

Astrolabe Astronomy Club,
  ✍ *www.neofoundation.org/astrolabe*
Central Appalachian Astronomy Club,
  ✍ *www.caacwv.org*
Kanawha Valley Astronomical Society,
  ✍ *www.kvas.org*

WISCONSIN

Madison Astronomical Society,
  ✍ *www.madisonastro.org*
Milwaukee Astronomical Society,
  ✍ *www.milwaukeeastro.org*
Northeast Wisconsin Stargazers, ✍ *www.new-star.org*
Sheboygan Astronomical Society,
  ✍ *www.shebastro.org*

WYOMING

Cheyenne Astronomical Society,
  ✍ *http://users.sisna.com/mcurran*
Laramie Astronomical Society, ✍ *www.lariat.org/LASSO*

In Canada

Astro Club Borealis, ✍ *www.osco.nb.ca/nbanb/*
Barrie Astronomy Club, ✍ *www.deepskies.com*
Big Sky Astronomical Society, ✍ *www.bigsky.ab.ca*
Club des Astronomes Amateurs de Longueuil,
  ✍ *www.aei.ca/~caal/*
Club des Astronomes Amateurs de Sherbrooke,
  ✍ *www.caas.sherbrooke.qc.ca*
Fraser Valley Astronomer's Society, ✍ *www.fvas.net*
Hamilton Amateur Astronomers,
  ✍ *www.science.mcmaster.ca/HAA*
La société d'astronomie de Montréal,
  ✍ *www.cam.org/~sam*
Le Club d'Astronomie de Rimouski,
  ✍ *www.quebectel.com/astro*
North York Astronomical Association,
  ✍ *www.nyaa-starfest.com*
Prince George Astronomical Society,
  ✍ *www.pgweb.com/~astronomical*
Royal Astronomical Society of Canada:
    Edmonton, ✍ *www.edmontonrasc.com*
    Halifax, ✍ *http://halifax.rasc.ca*
    London, ✍ *http://phobos.astro.uwo.ca/~rasc*
    Ottawa, ✍ *http://ottawa.rasc.ca*
York-Simcoe Amateur Astronomers Club,
  ✍ *http://hometown.aol.com/tdcarls/welcome.html*

Museum/Planetarium/ Observatory Listings
In the United States

ALABAMA
W.A. Gayle Planetarium, Troy State University,
 www.tsum.edu/planet

ALASKA
Imaginarium, *www.imaginarium.org*

ARIZONA
Flandrau Science Center, *www.flandrau.org*
Fred Lawrence Whipple Observatory,
 http://cfa-www.harvard.edu/flwo
Kitt Peak National Observatory,
 www.noao.edu/kpno
Lowell Observatory, *www.lowell.edu*
Mount Graham Observatory,
 http://mgpc3.as.arizona.edu
Steward Observatory, *www.as.arizona.edu*

ARKANSAS
University of Arkansas/Little Rock Planetarium,
 www.physics.ualr.edu/planetarium.html
University of Central Arkansas Planetarium,
 www.uca.edu/divisions/academic/physics/ Astronomy/UCA_Planetarium.html

CALIFORNIA
Big Bear Lake Solar Observatory,
 www.bbso.njit.edu
Exploratorium, *www.exploratorium.edu*
Griffith Observatory, *www.griffithobs.org*
Hat Creek Radio Observatory,
 http://bima.astro.umd.edu/general/hatcreek location.html
Lawrence Hall of Science, *www.lhs.berkeley.edu*
Lick Observatory, *www.ucolick.org*
Millikan Planetarium,
 www.astronomy.pomona.edu/millikan.html

Mountain Skies Astronomical Society,
 www.mountain-skies.org
Palomar Observatory,
 www.astro.caltech.edu/observatories/palomar
Reuben H. Fleet Science Center, *www.rhfleet.org*
Roth Planetarium,
 http://rigel.csuchico.edu/roth/roth.html
San Diego Aerospace Museum,
 www.aerospacemuseum.org

COLORADO
Black Forest Observatory, *www.observatory.org*
Fiske Planetarium, *www.colorado.edu/fiske*
Las Brisas Observatory, *www.lbo.teuton.org*
Star Chaser Observatory, *www.starchaser-obs.com*

CONNECTICUT
Gengras Planetarium, *www.sciencecenterct.org*
Talcott Mountain Space Center,
 www.sciencecenterct.org

DELAWARE
Mount Cuba Astronomical Observatory,
 www.physics.udel.edu/MCAO

DISTRICT OF COLUMBIA
Explorer's Hall,
 www.nationalgeographic.com/explorer
National Air and Space Museum, *www.nasm.edu*
United States Naval Observatory,
 www.usno.navy.mil

FLORIDA
Aldrin Planetarium, *www.sfsm.org*
Astronaut Hall of Fame, *www.astronauts.org*
Bishop Planetarium, *www.sfmbp.org*
Calusa Nature Center, *http://calusanature.com*
Kennedy Space Center,
 www.kennedyspacecenter.com
Moore Observatory, *www.mooreobservatory.com*
Orlando Science Center, *www.osc.org*

GEORGIA
Fernbank Science Center, &*www.fernbank.edu*
Rollins Planetarium, &*www.yhc.edu/planet*
University of Georgia Observatory,
&*www.physast.uga.edu/~star*

HAWAII
Bishop Museum Planetarium,
&*www.bishopmuseum.org*
Canada-France-Hawaii Telescope,
&*www.cfht.hawaii.edu*
Mauna Kea Observatory,
&*www.ifa.hawaii.edu/mko*

IDAHO
Faulkner Planetarium,
&*www.csi.edu/support/museum/faulkner/*
welcome.html

ILLINOIS
Adler Planetarium, &*www.adlerplanetarium.org*
Dearborn Observatory, &*www.astro.nwu.edu*
Discovery Center Museum,
&*www.discoverycentermuseum.org*
Illinois State University Planetarium,
&*www.phy.ilstu.edu/~wenning/planet.html*
John Deere Planetarium,
&*http://helios.augustana.edu/astronomy*
Lakeview Museum Planetarium,
&*www.lakeview-museum.org*
Staerkel Planetarium,
&*www.parkland.cc.il.us/coned/pla*
Strickler Planetarium,
&*http://website.olivet.edu/planetarium*

INDIANA
Ball State University Planetarium,
&*www.bsu.edu/physics/astro/Astronomy.html*
Kirkwood Observatory,
&*www.astro.indiana.edu/kirkwood_*
observatory.html

Koch Planetarium,
&*www.emuseum.org/shows.html*
Valaparaiso University Planetarium,
&*www.physics.valpo.edu*

IOWA
Menke Observatory,
&*http://electro.sau.edu/Homepage/*
Observatory.html
Science Center of Iowa, &*www.sciowa.org*

KANSAS
Kansas Cosmosphere and Space Center,
&*www.cosmo.org*
Lake Afton Public Observatory,
&*http://webs.wichita.edu/lapo*

KENTUCKY
Louisville Science Center, &*http://lsclouisnet.org*
Rauch Planetarium,
&*www.louisville.edu/planetarium*
Weatherford Observatory,
&*http://physics.berea.edu/observatory.htm*

LOUISIANA
Highland Road Park Observatory,
&*www.phys.lsu.edu/observatory*
Shreveport Planetarium, &*www.abcstar.com*

MAINE
Jordan Planetarium,
&*www.ume.maine.edu/~lookup*
Southworth Planetarium,
&*www.usm.maine.edu/planet*

MARYLAND
Crosby Ramsey Memorial Observatory,
&*www.mdsci.org/obs.htm*
Montgomery College Planetarium,
&*www.mc.cc.md.us/Departments/planet*

NASA-Goddard Space Flight Center,
   http://pao.gsfc.nasa.gov/vc/vc.htm
University of Maryland Observatory,
   www.astro.umd.edu/openhouse

MASSACHUSETTS

Boston Museum of Science, *www.mos.org*
Harvard-Smithsonian Center for Astrophysics,
   http://cfa-www.harvard.edu
Hopkins Observatory,
   www.williams.edu/Astronomy/Hopkins
Norton Observatory, *www.ecotarium.org*
Oak Ridge Observatory, *http://cfa-*
   www.harvard.edu/cfa/oir/OakRidge/
   oak.ridge.shtml

MICHIGAN

Abrams Planetarium, *www.pa.msu.edu/abrams*
Chaffee Planetarium,
   www.grmuseum.org/chaffeeplanetarium/
   chaffee.htm
Detroit Science Center, *www.sciencedetroit.org*
Exhibit Museum Planetarium,
   www.exhibits.lsa.umich.edu
Fox Observatory, *www.foxobservatory.org*
Michigan State University Observatory,
   www.pa.msu.edu/astro/observ
Sherzer Observatory,
   www.physics.emich.edu/sherzer

MINNESOTA

Alworth Planetarium, *www.d.umn.edu/~planet*
Onan Observatory, *www.mnastro.org/onan*
Paul Feder Observatory,
   www.moorhead.msus.edu/regsci
Science Museum of Minnesota, *www.smm.org*

MISSOURI

Kansas City Museum, *www.kcmuseum.com*
St. Louis Science Center, *www.slsc.org*

NEBRASKA

Kountze Planetarium,
   www.physics.unomaha.edu/planet
Mueller Planetarium, *www.spacelaser.com*

NEVADA

Fleischmann Planetarium,
   http://planetarium.unr.nevada.edu

NEW HAMPSHIRE

Christa McAuliffe Planetarium, *www.starhop.com*
University of New Hampshire Observatory,
   http://pubpages.unh.edu/~cbsiren/
   observatory.html

NEW JERSEY

Dreyfuss Planetarium,
   www.newarkmuseum.org/planetarium
Liberty Science Center, *www.lsc.org*
Novins Planetarium, *www.ocean.cc.nj.us/planet*

NEW MEXICO

Apache Point Observatory, *www.apo.nmsu.edu*
National Solar Observatory,
   www.sunspot.noao.edu
VLA: National Radio Astronomy Observatory,
   www.nrao.edu/

NEW YORK

Custer Observatory, *www.custerobservatory.org*
Hayden Planetarium,
   www.amnh.org/rose/haydenplanetarium.html
Kopernik Space Education Center,
   www.kopernik.org
Mount Stony Brook Observatory,
   http://ess.sunysb.edu/observer/mtsb.html
Strasenburgh Planetarium, *www.rmsc.org*
Vanderbilt Planetarium,
   www.vanderbiltmuseum.org/
   planetarium.htm

NORTH CAROLINA
Morehead Planetarium, &www.morehead.unc.edu
Schiele Museum of Natural History,
&www.schielemuseum.org

OHIO
Bowling Green State University Planetarium,
&http://physics.bgsu.edu/planetarium
Cleveland Museum of Natural History,
&www.cmnh.org
McKinley Museum and Planetarium,
&www.mckinleymuseum.org
Ohio Center of Science and Industry,
&http://cosi.org
Ohio State University Planetarium,
&www.astronomy.ohio-state.edu/
planetarium.html
Perkins Observatory, &www.perkins-observatory.org
Warner and Swasey Observatory,
&http://astrwww.cwru.edu
Warren Rupp Observatory,
&www.wro.org/home.htm

OKLAHOMA:
Omniplex, &www.omniplex.org
RMCC Observatory,
&http://astrotulsa.com/club/observatory.asp

OREGON
Mount Hood Community College Planetarium,
&www.starstuff.com/stars.htm
Oregon Museum of Science and Industry,
&www.omsi.edu

PENNSYLVANIA
Allegheny Observatory, &www.pitt.edu/~aobsvtry
Allentown School District Planetarium,
&http://asd.planetarium.org
Carnegie Science Center,
&www.carnegiesciencecenter.org

University of Pennsylvania Observatory,
&http://observatory.astro.upenn.edu

RHODE ISLAND
Cormack Planetarium, &www.osfn.org/museum
Frosty Dew Observatory,
&www.frostydrew.org/observatory

SOUTH CAROLINA
Dooley Planetarium,
&http://astro.fmarion.edu/planet
Francis Marion University Observatory,
&http://astro.fmarion.edu/observe
Melton Memorial Observatory,
&http://astro.physics.sc.edu/htmlpages/
melton/melton.html
Roper Mountain Observatory,
&www.ropermountain.org/index.shtml

SOUTH DAKOTA
Badlands Observatory,
&www.sdsmt.edu/space/bo.htm

TENNESSEE
Jones Observatory, &www.utc.edu/~jonesobs
Sharpe Planetarium,
&www.memphismuseums.org/planet.htm
Sudekum Planetarium,
&www.sudekumplanetarium.com

TEXAS
Austin State University Planetarium,
&www.physics.sfasu.edu/planetarium
Hudnall Planetarium,
&www.tyler.cc.tx.us/planet/planet.htm
McDonald Observatory, www.as.utexas.edu
Scobee Planetarium,
&www.accd.edu/sac/ce/scobee
Sky Theater Planetarium,
&www.phys.unt.edu/planetarium/index.htm
Space Center Houston, &www.spacecenter.org

UTAH

Hansen Planetarium, ✍ *www.hansenplanetarium.org*
Ott Planetarium,
✍ *http://physics.weber.edu/planet/ott.html*

VERMONT

Fairbanks Museum and Planetarium,
✍ *www.fairbanksmuseum.com*

VIRGINIA

Hopkins Planetarium,
✍ *www.smwv.org/planetarium.htm*
Keeble Observatory,
✍ *www.rmc.edu/academics/phys/keeble*
University of Virginia Observatories,
✍ *www.astro.virginia.edu/research/
observatories*
Virginia Air and Space Center, ✍ *www.vasc.org*

WASHINGTON

Western Washington University Planetarium,
✍ *www.ac.wwu.edu/~skywise*

WEST VIRGINIA

National Radio Astronomy Observatory,
Green Bank, ✍ *www.gb.nrao.edu*
Sunrise Museum, ✍ *www.sunrisemuseum.org*
Tomchin Planetarium, ✍ *www.as.wvu.edu/~planet*

WISCONSIN

Barlow Planetarium, ✍ *www.fox.uwc.edu/barlow*
Milwaukee Public Museum, ✍ *www.mpm.edu*
Washburn Observatory,
✍ *www.astro.wisc.edu/Washburn*
Yerkes Observatory,
✍ *http://astro.uchicago.edu/yerkes/*

WYOMING

Casper Planetarium,
✍ *www.trib.com/WYOMING/NCSD/
PLANETARIUM/planetarium.html*

In Canada

ASTROLab du Mont-Megantic,
✍ *astrolab.interlinx.qc.ca*
Burke-Gaffney Observatory,
✍ *http://www.ap.stmarys.ca/bgo*
Calgary Science Centre, ✍ *www.calgaryscience.ca*
Doran Planetarium,
✍ *http://laurentian.ca/physics/Planetarium/
planetarium.html*
Edmonton Space and Science Centre,
✍ *www.ee.ualberta.ca/essc*
MacMillan Space Centre,
✍ *www.hrmacmillanspacecentre.com*
Montreal Planetarium,
✍ *www.planetarium.montreal.qc.ca*
Ontario Science Center,
✍ *www.ontariosciencecentre.ca*
Science North Solar Observatory,
✍ *sciencenorth.on.ca*
Science World British Columbia,
✍ *www.scienceworld.bc.ca*

Glossary

A

aborigine: From Latin, "people who have been present from the beginning"; When capitalized refers to indigenous Australians.

absolute magnitude: The brightness of a star as seen from a standard distance, defined as 10 parsecs.

accretion: A gradual growth or increase in size due to a slow buildup of material.

active galactic nucleus: The core of an active galaxy; usually abbreviated AGN.

active galaxy: A type of galaxy not defined by its shape or structure; these galaxies have a core, or nucleus (the active galactic nucleus), that is thought to generate huge amounts of energy.

afocal photography: A method of astrophotography that places the camera near the telescope eyepiece with both lenses focused at infinity.

albedo: The fraction of sunlight that is reflected off the surface of an object in the solar system; objects with high albedo (near 1) are bright, while objects with low albedo (near 0) are dark.

alt-az mount: Altitude-azimuth mount is a type of telescope mount involving axes that will move both horizontally and vertically.

altitude: The vertical height or elevation of an object, measured from the horizon or an arbitrary plane.

ante meridiem: Latin for "before meridian." Abbreviated A.M., it refers to the time of day when the Sun is east of the meridian, i.e., the morning.

aperture: An opening designed to let light through; the lens diameter in a telescope.

aphelion: The point in the orbit of a planet (or other celestial body) where it is farthest from the Sun

apparent magnitude: The brightness of a star as seen from Earth.

archeoastronomy: A combination of archeology and astronomy that involves the study of religion, folklore, celestial myths, and all ancient astronomical rituals and ideas.

armillary: A spherical astronomical instrument used to measure the longitude and latitude of planets and stars.

asterism: A small group of stars that is not a constellation.

asteroid: A rocky or metallic, atmosphere-free body that orbits around the Sun, usually less than 1,000 kilometers in diameter.

astrolabe: An instrument using spherical projections to locate the position of celestial bodies.

astrology: The study of how the movements of the stars and planets affect us and our lives.

astronomical compendium: A small measuring device used during the Renaissance; could contain map, sundial, compass, astrolabe, sextant, and other astronomical tools.

astronomy: The study of stars, planets, and other bodies outside Earth's atmosphere.

astrophotography: Photography used in astronomy; can involve photography through telescopes.

atmosphere: The gaseous material surrounding planets; the air surrounding Earth.

azimuth: The direction of an object (planet, star, or other object in the cosmos) measured around the horizon clockwise.

B

barred spiral galaxy: A type of galaxy with arms of stars that spiral out from a linear central bar.

Big Bang theory: The scientific explanation for the origin of the universe, championing the notion that the universe spontaneously originated 15 billion years ago in a colossal explosion

binary star: A two-star system, where the individual stars orbit their common center of mass, often appearing to revolve around each other.

black hole: A location in space with very high mass and density; gravity prevents anything, including light, from escaping such a region.

blazar: A very energetic quasar that emits electromagnetic radiation.

C

cairn: a pile of stones used as a landmark, usually covering an underground chamber.

cataclysmic variable star: A type of variable star whose brightness is altered suddenly and violently by occasional explosions from within the star.

catadioptric telescope: A telescope that uses both mirrors and lenses to reflect and gather light; a compound telescope.

CBAT: Central Bureau for Astronomical Telegrams.

celestial sphere: An imaginary hollow sphere surrounding Earth, with all the stars painted on the inside.

charged coupled device: A computer chip with a grid of embedded light-sensitive detectors; (CCD).

closed universe: The theory that if the density of the universe is greater than the critical density, expansion will eventually cease, and the universe will someday collapse into to a single point.

codex: A manuscript book.

collimation: Creating parallel axes, or alignment of binocular or telescope optics to their mechanical axes.

comet: An ice-rock body composed of ice and nonvolatile dust, constituted mainly of frozen water or gas. Comets typically have five parts: a nucleus, a hydrogen cloud, a coma, a dust tail, and an ion tail.

constellation: A configuration of stars.

convection: The transfer of heat through the circulation of heated parts of liquid or gas.

convective zone: The top 15 percent of the Sun, where convection takes place.

coordinate system: The indexing of two or more terms (or coordinates) so that the intersection of their points is a meaningful way to locate objects.

cosmology: The study of the origins of the universe.

crater: A hole in the ground, usually hemispherical, resulting from the impact of a heavy, falling object.

critical density: The density of the universe at which the fate of the universe changes from one that expands forever (open universe) to one that will eventually stop expanding and begin contracting (closed universe).

D

dark matter: Material that does not give off any radiation, and is suspected of making up a sizeable fraction of the universe.

dark nebula: A cloud of gas and dust that is large and dense enough to block out the light given off by something behind it.

declination: Similar to latitude it is the angular distance to a celestial object, measured from the celestial equator; it helps identify the positions of stars and other celestial bodies.

Dobsonian mount: A type of telescope mount that uses a simplified altitude-azimuth design; the telescope maintains position by the friction created between the telescope bearings and the mount. Promoted by John Dobson of the Sidewalk Astronomers.

doppler effect: The observation that the pitch of a sound changes when it moves toward or away from an observer.

dualism: The sixteenth-century philosophical notion that mind and matter are completely separate substances, but mutually affected by God.

E

eccentric: Elliptical, deviating from circular.

eclipse: The passing of one celestial body in front of another.

eclipsing binary star: A type of binary star system whose orbital plane lies close enough to our line of sight that it appears to eclipse itself, appearing as a variable star.

ejecta: Material that has been thrown out, or ejected, from something, usually material from an impact crater.

elliptical galaxy: A type of galaxy that is shaped like an ellipse.

emission nebula: A very hot gas cloud; light from nearby stars causes these nebulae to emit radiation as transitions take place from high-energy to low-energy states.

epicycle: A circle in which a planet moves, while at the same time rotating around its own center. Required in geocentric views of the solar system.

equinox: One of two times each year when the Sun crosses the equator, so day and night are equally long—generally, March 21 and September 23. Also, the place on the celestial sphere where the ecliptic crosses the celestial equator.

erosion: The state of being diminished or reduced over time, usually through repeated application of natural elements such as water, wind, or ice.

escape velocity: The speed at which something needs to travel in order to escape the gravitational pull of a star or planet.

event horizon: The distance from the center of a black hole inside which everything approaching the black hole falls inward toward the singularity.

exit pupil: The ratio of the diameter of the objective lens to the binoculars' magnifying power.

extrasolar: A descriptive term for planets that are not in our solar system.

extraterrestrial: Occurring outside the atmosphere and boundaries of Earth.

eye relief: The distance the viewer's eyes need to be from the binocular eyepiece in order to take in the entire field of view.

F

finder scope: A small telescope that attaches to a larger one; allows you to align your view on certain stars or planets.

fireballs: Meteors that appear brighter than anything in the sky except for the Sun and Moon.

focal length: In optics, refers to the distance from the center of a lens to the point where it is in focus. In a telescope, this is usually the length of the tube.

G

galactic cluster: A grouping of stars that is less dense than a globular cluster; an irregular mass, usually with no central core, generally having no more than a few thousand stars.

galaxy: A large group of stars, gasses, and dust held together by gravitation and separated from similar systems by vast regions of space.

gas giant: A large planet, such as Jupiter, that is composed mostly of gaseous elements instead of rocky material.

geocentric: A view that Earth is at the center of the solar system.

German equatorial mount: A telescope mount designed to follow the movements of planets or stars across the sky.

globular cluster: A very dense grouping of stars existing inside or next to a galaxy, containing anywhere from 100,000 to 1 million stars.

gravitational lensing: A phenomenon that occurs when an object has enough mass that it can bend light passing by it from an object behind it, causing observers on Earth to receive multiple images of the same object.

greenhouse effect: The gradual raising of the temperature of Earth's surface and atmosphere, due to a buildup of carbon dioxide and water vapor which traps the radiation from the Sun.

H

habitable zone: The area surrounding a star at a distance from the star where the temperature is between the freezing and boiling points of water.

heliocentric: A view that the Sun is at the center of the solar system.

hemisphere: Half of a sphere.

hydrothermal vent: A fissure or crack in the bottom of an ocean through which hot liquids or other materials pass up from the planet's interior

I

intelligent life: Generally considered to be life forms that can communicate with humans.

irregular galaxy: A type of galaxy that is not arranged in any orderly pattern and has no observable symmetry, such as a spiral or ellipse.

K

kiva: Partially underground chambers used for spiritual rituals by Native Americans.

L

latitude: The location of a point on Earth's surface north or south of the equator, expressed in degrees.

lens: A piece of glass or plastic having either curved or straight surfaces; used in optical instruments to focus light rays.

longitude: the number of degrees a body is located east or west of an arbitrarily defined starting point (usually the prime meridian) on Earth's surface.

M

magnetic field: A condition generated by electrons moving through space.

magnification: The state of apparent enlargement, especially when something is viewed through an optical tool such as a telescope.

magnitude: In the context of astronomy, the brightness of a star as seen from Earth; the brighter a star, the smaller its magnitude.

maria: Latin for ocean, refers to dark patches on the Moon, many of which are ancient lava flows.

medicine wheel: A rock arrangement closely resembling a spoked wheel, used by Native Americans.

megalith: A large, rough stone used for buildings or monuments in prehistoric times.

megaparsec: A unit of measurement for distance in space; equals 1 million parsecs.

meridien: an imaginary line on the Earth or the celestial sphere that goes from pole to pole through the observer's location (or a point directly above it).

Messier's catalogue: A description of 110 galaxies and other nonstar objects as observed and documented by Charles Messier in 1774.

meteor: Seen as a bright streak in the sky (sometimes called a *shooting star* or *falling star*) it is a tiny grain of dust or pebble that burns up when it hits Earth's atmosphere. Meteors come from comets and other space debris.

meteor shower: A regular occurrence where multiple meteors are visible.

meteorite: A meteor that survives passage through the atmosphere and crashes into Earth.

meteoroid: Small piece of rock or dust floating freely in space, before they enter the Earth's atmosphere and burn up as meteorites.

moon: A natural satellite of a planet.

mutation: A permanent change, either physical or chemical.

N

NASA: National Aeronautics and Space Administration. Founded in 1958, the American space agency has been responsible for a number of remarkable breakthroughs in aeronautics and space exploration.

near-Earth asteroids: Asteroids that come within approximately 120 million miles of the Sun.

nebula: A large cloud consisting of dust and gas that exists in interstellar space.

neutron star: A very dense core of material that forms after a star explodes.

nova: A star that has a sudden bright period, then a sharp decline in brightness.

nuclear fusion: The process by which the nuclei of two atoms fuse together to form a heavier element; large amounts of energy are produced by this fusion.

nucleosynthesis: The process by which hydrogen nuclei combine in pairs to form helium nuclei, and so on to form the elements in the early universe and inside stars.

O

observatory: A building or location equipped to watch astronomical phenomena.

occultation: In astronomy, the occurrence of one large celestial body moving in front of another, smaller one.

orbit: The path one object takes as it revolves around another, usually due to external forces such as gravity.

orbital resonance: The condition resulting from a two-planet or satellite system where one orbital period is twice that of the other. Resonances are possible with more than two planets; three of the Galilean satellites of Jupiter are in 1:2:4 resonances with each other.

organism: A collection of interdependent parts that survives and thrives by acting as a single unit.

origin: The point at which things begin; zero.

P

parapegma: A stone tablet that allowed Greek astronomers to make connections between dates and planetary movements, a very early predecessor to the planisphere.

parsec: A unit of measurement for distances in space that equals 3.26 light-years. It's defined as the distance at which the radius of Earth's orbit around the Sun would measure 1 arc-second.

perihelion: The point in the orbit of a planet (or other solar system object) where it is nearest the Sun.

photosphere: The surface of the Sun, the deepest layer that can be visibly observed.

planet: A large body that revolves around a star, such as the Sun in our solar system, and is not sufficiently massive to ignite and undergo nuclear fusion.

planetarium: A room, usually in a science museum or observatory, where celestial objects and events are projected onto the walls and ceiling.

planetary nebula: Gas ejected from a star that is approaching the end of its life; resembled a planet to early telescopic observers.

planisphere: The projection of a sphere onto a flat surface, or plane. In astronomy, a projection of the celestial skies into a plane; also called a *star wheel*.

plume: A long, rising column of gas or smoke.

post meridiem: Latin for "after meridian." Abbreviated P.M., refers to the time of day when the Sun is west of the meridian, i.e., the afternoon.

prime meridian: The zero line for longitude at the British Royal Greenwich Observatory, in Greenwich England.

prism: A transparent object that disperses light and separates it into the individual colors of the spectrum.

prograde motion: The regular movement of planets in the sky, from west to east; opposite of retrograde, which is east to west.

protostar: A clump in a cloud of gas which heats and collapses, but is not yet a true star because it does not perform nuclear fusion.

pulsar: A neutron star that spins extremely rapidly.

pulsating variable star: A type of variable star in which the actual layers of the star expand and contract, giving off varying amounts of light.

Q

quadrant: An instrument consisting of a graduated arc with a plumb line, used to measure the altitude of celestial bodies.

quasar: Means *quasi-stellar object*; a class of celestial objects that look like stars when photographed, but have high redshifts and give off strong radio emissions.

R

radiation: The emission of energy as electromagnetic waves.

radiative zone: The region outside the core of the Sun where energy travels from the interior to outer layers.

radio galaxy: A galaxy that gives off radio waves.

red giant: A star near the end of its lifetime with a relatively low surface temperature and large diameter when compared to our Sun; the star is cool because its energy is spread out over such a large diameter.

redshift: The shifting of an object's spectrum toward longer wavelengths due to its motion away from an observer.

reflecting telescope: A telescope that uses a concave mirror to gather light.

reflection nebula: A dust and gas cloud that reflects the light from nearby stars.

refracting telescope: A telescope that gathers light through a lens, sending the light to the eyepiece.

Renaissance: The historical period in Europe between the fourteenth and seventeenth centuries that was characterized by a rebirth of interest in art, music, astronomy, humanism, and other cultural and sociological events.

retrograde motion: When internal and external planetary orbits coincide in such a manner that it appears the planet has reversed directions in the sky; movement from east to west; opposite of prograde motion.

right ascension: Similar to longitude; a measurement from 0 to 24 hours that helps identify the positions of stars and other celestial bodies.

S

satellite: A celestial body that orbits another, larger, celestial body that is not a star; also called a *natural satellite*, as distinct from a robotic *artificial satellite*.

SETI: Search for Extraterrestrial Intelligence; also the SETI Institute, a research institution located in Mountain View, California.

sextant: An instrument used to measure the angular distance between stars or planets.

Seyfert galaxy: A type of active (spiral) galaxy that emits low-energy gamma rays.

silicate: A common rocky material containing silicon, oxygen, and one or more metals; the primary component of some types of asteroids as well as many other solar system bodies.

singularity: In the Big Bang theory, the point where time becomes zero and the density goes to infinity. Also can refer to the point at the center of a black hole.

solstice: One of two points on the ecliptic where the distance from the equator is at its maximum; summer solstice occurs around June 22, winter solstice around December 22. These are the longest and shortest days in the Northern Hemisphere.

spacecraft: A man-made vehicle designed to pass through Earth's atmosphere into orbit or beyond.

spectroscopy: The process of passing light through a prism to study different wavelengths.

spectrum: The color band created when white light is scattered.

spiral galaxy: A type of galaxy with arms of stars that appear to spiral out from a central core.

star: A self-luminating ball of gas that produces energy from internal nuclear reactions.

star party: A gathering held in a dark-sky location, where amateur astronomers bring telescopes and get together to observe celestial objects and events.

steady state theory: The notion that the universe has been around forever and is slowly expanding.

subatomic: Pertains to the particles contained within an atom, such as electrons, neutrons, or protons; smaller than an atom

subsurface: Means beneath the surface layer.

supernova: An explosion at the end of the life of a large star; it can be up to a billion times brighter than the Sun.

T

terrestrial: Relating to Earth.

theodolite: An instrument used by surveyors and astronomers for measuring both altitude and azimuth.

tidal flexing: A condition that occurs when the gravitational pull of a planet tugs on one of its satellites, causing the satellite's crust to flex up and down.

tidal heating: The warming of a planet due to tidal flexing.

transit: In astronomy, the occurrence of something small in our solar system moving directly in front of something larger.

U

universe: Everything that exists; the cosmos.

V

variable stars: Stars that change brightness periodically.

W

wavelength: The distance between corresponding peaks in different phases of a wave.

white dwarf: A small, very dense star object that cannot collapse into itself any further; one possible result for a low-mass star at the end of its life.

white hole: The theoretical "other end" of a black hole.

Worm hole: The theoretical connection between a black hole and a white hole.

Z

Zodiac: Greek for "circle of animals," a group of twelve symbols and mythological stories that correspond to twelve constellations.

PHOTOGRAPHY CAPTIONS

FIGURE 0-1: The universe is much larger than it appears from Earth. This spiral galaxy shows how our own galaxy might appear from a distance.
Image courtesy of NASA and The Hubble Heritage Team (Association of Universities for Research in Astronomy [AURA], Space Telescope Science Institute [STScI], NASA)

FIGURE 0-2: The Globular Cluster M15
Image courtesy of NASA and The Hubble Heritage Team (Association of Universities for Research in Astronomy [AURA], Space Telescope Science Institute [STScI], NASA)

FIGURE 1-1: Timeline of ancient astronomy 50,000 B.C. to 5000 B.C. *Created by Shana Priwer*

FIGURE 1-2: Timeline of ancient astronomy 3,000 B.C. to 2000 A.D. *Created by Shana Priwer*

FIGURE 1-3: Stonehenge was one of the world's first physical calendars. *© 2001, www.arttoday.com*

FIGURE 1-4: The arrangement of these Egyptian pyramids is said to correspond to the alignment of the stars on Orion's belt. *Created by Shana Priwer*

FIGURE 2-1: Geocentrism and heliocentrism were, literally, worlds apart both in conception and reality.
Created by Shana Priwer

FIGURE 3-1: The Big Bang timeline shows when scientists believe different parts of the universe were formed.
Created by Shana Priwer

FIGURE 3-2: Clouds of dust and gas such as the Keyhole Nebula provide the raw materials for the initial formation of stars.
Image courtesy of NASA and The Hubble Heritage Team (Association of Universities for Research in Astronomy [AURA], Space Telescope Science Institute [STScI], NASA)

FIGURE 3-3: An early picture of what was once called the Andromeda Nebula. We now know that the Andromeda Galaxy is 900,000 light-years from us!
© 2001, www.arttoday.com

FIGURE 3-4: Galaxy NGC 4214. The same star formation process that began in the Big Bang continues today in galaxies spread across the universe.
Image courtesy of NASA and The Hubble Heritage Team (Association of Universities for Research in Astronomy [AURA], Space Telescope Science Institute [STScI], NASA)

FIGURE 5-1: Latitude spans the distance from 0 degrees at the Equator to 90 degrees at the Poles.
© 2001, www.arttoday.com

FIGURE 5-2: Globes show lines of latitude and longitude, which are extremely useful for locating places on Earth.
© 2001, www.arttoday.com

FIGURE 5-3: The celestial sphere makes a convenient visual analogy for thinking about how the heavens seem to move above us on Earth. *Created by Shana Priwer*

FIGURE 5-4: A long exposure of the night sky centered on Polaris, the North Star, tracks the circular motions of the stars around the north celestial pole.
© 2001, www.arttoday.com

FIGURE 6-1: The planet Mercury, from the Mariner 10 spacecraft. *Courtesy of NASA/Jet Propulsion Laboratory (JPL), Caltech*

FIGURE 6-2: Combined radar/topography view of the surface of the planet Venus, from the *Magellan* spacecraft. Vertical heights have been exaggerated.
Courtesy of NASA/Jet Propulsion Laboratory (JPL), Caltech

FIGURE 6-3: The planet Earth, as seen from the *Apollo 17* spacecraft. *Courtesy of NASA/Johnson Space Center [JSC]*

FIGURE 6-4: A close-up view of the Moon's surface from orbit, taken from *Apollo 10*.
Courtesy of NASA/Johnson Space Center [JSC]

FIGURE 6-5: This mountain on Mars appeared to be a face in early low-resolution images. New pictures from the *Mars Global Surveyor* spacecraft, however, have shown that it is just a natural mountainous feature, and not an artifact made by intelligent life.
Courtesy of NASA/Jet Propulsion Laboratory (JPL), Caltech

FIGURE 6-6: The surface of Mars. *Courtesy of NASA, JPL, Caltech*

FIGURE 7-1: The planet Jupiter.
Courtesy of NASA/Jet Propulsion Laboratory, (JPL), Caltech

FIGURE 7-2: The four Galilean satellites as seen from the *Galileo* spacecraft. From left: Io, Europa, Ganymede, and Callisto.
Courtesy of NASA/Jet Propulsion Laboratory (JPL), Caltech

FIGURE 7-3: The planet Saturn.
Image courtesy of NASA and The Hubble Heritage Team (Association of Universities for Research in Astronomy [AURA], Space Telescope Science Institute [STScI], NASA)

FIGURE 7-4: The planet Uranus, with an exaggerated view at the right to show slight differences in cloud layers. The south pole is located near the middle of the disk.
Courtesy of NASA/Jet Propulsion Laboratory (JPL), Caltech

FIGURE 7-5: The planet Neptune.
Voyager image courtesy of NASA/Jet Propulsion Laboratory (JPL), Caltech

FIGURE 7-6: Our best view of the planet Pluto and its moon Charon.
Image courtesy of NASA and The Hubble Heritage Team (Association of Universities for Research in Astronomy [AURA], Space Telescope Science Institute [STScI], NASA) and Dr. R. Albrecht, European Space Agency (ESA)/European Southern Observatory (ESO), Space Telescope European Coordinating Facility.

PHOTOGRAPHY CAPTIONS

FIGURE 8-1: The northern hemisphere of Eros, taken by the *NEAR* spacecraft.
Courtesy of NASA/Jet Propulsion Laboratory (JPL), Caltech

FIGURE 8-2: The nucleus of comet Halley, taken from 20,000 kilometers by the *Giotto* spacecraft.
Image courtesy of NASA/National Space Science Data Center (NSSDC), Giotto.

FIGURE 8-3: Occasionally meteor showers can fill the sky with shooting stars. This occurrence is called a *meteor storm*, and is shown in this early artwork.
© 2001, www.arttoday.com

FIGURE 8-4: This image shows a small piece of rock burning up in Earth's atmosphere. A meteor of this size is called a *fireball*.
© 2001, www.arttoday.com

FIGURE 8-5: Meteor Crater is one of the largest meteor artifacts on Earth. © 2001, *www.arttoday.com*

FIGURE 9-1: Our Sun, the closest star to Earth, occasionally ejects material from its surface in a event called a *solar flare*. The energy given off in a solar flare can affect communication on Earth and cause auroras.
Courtesy of NASA/Jet Propulsion Laboratory (JPL), Caltech

FIGURE 9-2: Stars form as gas and dust from a nebula condensing into larger and larger clumps, which eventually ignite.
Image courtesy of NASA and The Hubble Heritage Team (Association of Universities for Research in Astronomy [AURA], Space Telescope Science Institute [STScI], NASA)

FIGURE 9-3: The Ant Nebula. This planetary nebula, numbered Mz 3, shows the last stage in the evolution of a star like the Sun. The nebula will last for only a few thousand more years during which time the gas will slowly disperse and the small stellar remnant in the middle will cool down to form a white dwarf.
Image courtesy of NASA and The Hubble Heritage Team (Association of Universities for Research in Astronomy [AURA], Space Telescope Science Institute [STScI], NASA)

FIGURE 10-1: Whirlpool Galaxy M-51. The well-defined structure of this spiral galaxy gives scientists extremely useful information about the formation and evolution of spiral galaxies.
Image courtesy of NASA and The Hubble Heritage Team (Association of Universities for Research in Astronomy [AURA], Space Telescope Science Institute [STScI], NASA)

FIGURE 10-2: Galaxy ESO 510-G13. The twisted aspect of this warped galaxy, probably caused by gravity distortions, makes it quite unusual.
Image courtesy of NASA and The Hubble Heritage Team (Association of Universities for Research in Astronomy [AURA], Space Telescope Science Institute [STScI], NASA)

FIGURE 10-3: Globular Cluster NGC-6093
Image courtesy of NASA and The Hubble Heritage Team (Association of Universities for Research in Astronomy [AURA], Space Telescope Science Institute [STScI], NASA)

FIGURE 10-4: Part of the constellation Lyra, the Ring Nebula is composed of gas cast off by a Sun-sized star near the end of its lifetime thousands of years ago.
Image courtesy of NASA and The Hubble Heritage Team (Association of Universities for Research in Astronomy [AURA], Space Telescope Science Institute [STScI], NASA)

FIGURE 10-5: The Horsehead Nebula's intricate structure, as revealed by the Hubble Space Telescope.
Image courtesy of NASA and The Hubble Heritage Team (Association of Universities for Research in Astronomy [AURA], Space Telescope Science Institute [STScI], NASA)

FIGURE 11-1: Galaxy M87. An extremely massive black hole lies at the center of this elliptical galaxy. Jets of electrons are emitted in a twisting magnetic field.
Image courtesy of NASA and The Hubble Heritage Team (Association of Universities for Research in Astronomy [AURA], Space Telescope Science Institute [STScI], NASA)

FIGURE 12-1: These binoculars are 7 x 50, meaning they have a magnification of seven- and fifty-millimeter lenses.
Image © 2001, www.arttoday.com

FIGURE 12-2: Many binoculars have a center wheel for focusing your view. *Image © 2001, www.arttoday.com*

FIGURE 13-1: Telescopes are complicated viewing instruments that can yield incredible images.
Image © 2001, www.arttoday.com

FIGURE 13-2: Setting up a telescope mount can be quite complicated—there are many factors to take into account, including declination and rotation.
Image © 2001, www.arttoday.com

FIGURE 14-1: Go to a star party! Share your telescope and learn from those around you.
Image © 2001, www.arttoday.com

FIGURE 14-2: Planetariums are a great place to learn about the night sky, especially if you live in a city or far from dark-sky locations. This image shows the Flandrau Planetarium at the University of Arizona in Tucson.
Image © 2001, www.arttoday.com

FIGURE 14-3: An early version of the Zeiss star projector, which is used in many planetariums to project moving images of the stars and planets onto a domed screen on the ceiling.
Image © 2001, www.arttoday.com

PHOTOGRAPHY CAPTIONS

FIGURE 14-4: Kitt Peak National Observatory, near Tucson, has a variety of telescopes including one with a four-meter mirror (center). Tours are available. *Image © 2001, www.arttoday.com*

FIGURE 14-5: Some museums display engineering mockups of space capsules similar to those capsules that actually went into space. This image was taken in the Smithsonian National Air and Space Museum in Washington, D.C.
Image © 2001, www.arttoday.com

FIGURE 15-1: Asteroid trail in Centarus. Note the trail in the upper-right corner.
Image courtesy of NASA and The Hubble Heritage Team (Association of Universities for Research in Astronomy [AURA], Space Telescope Science Institute [STScI], NASA)

FIGURE 15-2: Most newly discovered comets don't have significant tails like this one; instead, they just look like small bright fuzzy dots. The tail doesn't develop until the new comet comes substantially closer to the Sun.
Image © 2001, www.arttoday.com

FIGURE 15-3: This artist's conception of an aurora demonstrates the glow seen over water. *Image © 2001, www.arttoday.com*

FIGURE 15-4: This photograph shows the "diamond ring" phase of a total solar eclipse, which occurs just before the entire Sun is covered up by the Moon.
Image © 2001, www.arttoday.com

FIGURE 16-1: The electromagnetic spectrum usually goes from radio waves to x-rays. The visible spectrum lies close to the middle of the diagram. *Created by Shana Priwer*

FIGURE 16-2: This image shows both radio and optical telescopes on Kitt Peak, near Tucson, Arizona.
Image © 2001, www.arttoday.com

FIGURE 16-3: The dome covering this radio telescope at Kitt Peak, near Tucson, Arizona, opens during observations, and closes again when observations are complete. Similar to the arrangement in professional optical telescopes, the dome also rotates to allow the telescope to observe different parts of the sky.
Image © 2001, www.arttoday.com

FIGURE 16-4: Arecibo radio telescope. The large dish at the bottom sits in an old volcanic crater, and the signal feed suspended over the top collects the signals detected by the dish.
Image courtesy Taylor Bucci, 2001.

FIGURE 17-1: Venera 13, a Venus lander built by the former Soviet Union and the first spacecraft to return a color image from the surface of Venus.
Image courtesy of National Space Science Data Center (NSSDC)/NASA

FIGURE 17-2: The *Cassini* space probe, currently on its way to Saturn, undergoes preflight testing in a clean room at the Jet Propulsion Laboratory in California.
Image courtesy of National Space Science Data Center (NSSDC)/NASA

FIGURE 17-3: Edwin Aldrin descends onto the lunar surface, becoming the second human to set foot on the Moon.
Courtesy of NASA/Johnson Space Center [JSC]

FIGURE 17-4: A footprint on the surface of the Moon will last for thousands or millions of years.
Courtesy of NASA/Johnson Space Center [JSC]

FIGURE 17-5: The space shuttle *Columbia* coming in for a landing after its first flight into space.
Image courtesy of National Space Science Data Center (NSSDC)/NASA

FIGURE 17-6: Artist's conception of the finished International Space Station.
Image courtesy of National Space Science Data Center (NSSDC)/NASA

FIGURE 18-1: Habitable zones vary in size and location for different types of stars. *Created by Shana Priwer*

FIGURE 18-2: Planetary protection was very important after the lunar landings, and the first lunar astronauts were quarantined in this sealed facility following their return from the Moon.
Image courtesy of National Space Science Data Center (NSSDC)/NASA

FIGURE 18-3: This binary image was sent out by the Arecibo observatory in 1974. See *www.seti.org/science/a-message.html* for more information.
Image courtesy of NASA/Jet Propulsion Laboratory (JPL), Caltech

FIGURE 18-4: The *Pioneer 10* and *11* spacecraft both carried a simple pictorial plaque to show any intelligent life intercepting the spacecraft at some point in the distant future where *Pioneer* came from and who launched it.
Image courtesy of NASA

FIGURE 19-1: Planetary nebula NGC-3132. At the end of our Sun's lifetime, the Sun will briefly form a planetary nebula similar to this one, with a small white dwarf star in the middle.
Image courtesy of NASA and The Hubble Heritage Team (Association of Universities for Research in Astronomy [AURA], Space Telescope Science Institute [STScI], NASA)

FIGURE 19-2: Galaxies NGC 2207 and IC 2163. As the universe starts contracting, individual galaxies will begin to merge.
Image courtesy of NASA and The Hubble Heritage Team (Association of Universities for Research in Astronomy [AURA], Space Telescope Science Institute [STScI], NASA)

Index

THE EVERYTHING WEATHER BOOK

By Mark Cantrell

THE EVERYTHING WEATHER BOOK

From daily forecasts to blizzards, hurricanes, and tornadoes—all you need to know to be your own meteorologist

Mark Cantrell

Trade paperback, $14.95
1-58062-668-8, 304 pages

As an increase in severe weather phenomena has garnered media attention worldwide, the weather remains the single biggest topic of everyday conversation. *The Everything® Weather Book* provides readers with the perfect introduction to the complexities of weather, focusing on how weather develops, the causes of severe weather, the impact of global warming, and much more. Featuring dozens of photographs, *The Everything® Weather Book* builds readers' knowledge about tornadoes and hurricanes, rainbows, thunder and lightning, cloud formations, forecasting, and the Greenhouse Effect.

OTHER *EVERYTHING®* BOOKS BY ADAMS MEDIA CORPORATION

Everything® **After College Book**
$12.95, 1-55850-847-3

Everything® **American History Book**
$12.95, 1-58062-531-2

Everything® **Angels Book**
$12.95, 1-58062-398-0

Everything® **Anti-Aging Book**
$12.95, 1-58062-565-7

Everything® **Astrology Book**
$12.95, 1-58062-062-0

Everything® **Astronomy Book**
$14.95, 1-58062-723-4

Everything® **Baby Names Book**
$12.95, 1-55850-655-1

Everything® **Baby Shower Book**
$12.95, 1-58062-305-0

Everything® **Baby's First Food Book**
$12.95, 1-58062-512-6

Everything® **Baby's First Year Book**
$12.95, 1-58062-581-9

Everything® **Barbecue Cookbook**
$14.95, 1-58062-316-6

Everything® **Bartender's Book**
$9.95, 1-55850-536-9

Everything® **Bedtime Story Book**
$12.95, 1-58062-147-3

Everything® **Bible Stories Book**
$14.95, 1-58062-547-9

Everything® **Bicycle Book**
$12.00, 1-55850-706-X

Everything® **Breastfeeding Book**
$12.95, 1-58062-582-7

Everything® **Budgeting Book**
$14.95, 1-58062-786-2

Everything® **Build Your Own Home Page Book**
$12.95, 1-58062-339-5

Everything® **Business Planning Book**
$12.95, 1-58062-491-X

Everything® **Candlemaking Book**
$12.95, 1-58062-623-8

Everything® **Car Care Book**
$14.95, 1-58062-732-3

Everything® **Casino Gambling Book**
$12.95, 1-55850-762-0

Everything® **Cat Book**
$12.95, 1-55850-710-8

Everything® **Chocolate Cookbook**
$12.95, 1-58062-405-7

Everything® **Christmas Book**
$15.00, 1-55850-697-7

Everything® **Civil War Book**
$12.95, 1-58062-366-2

Everything® **Classical Mythology Book**
$12.95, 1-58062-653-X

Everything® **Coaching & Mentoring Book**
$14.95, 1-58062-730-7

Everything® **Collectibles Book**
$12.95, 1-58062-645-9

Everything® **College Survival Book**
$12.95, 1-55850-720-5

Everything® **Computer Book**
$12.95, 1-58062-401-4

Everything® **Cookbook**
$14.95, 1-58062-400-6

Everything® **Cover Letter Book**
$12.95, 1-58062-312-3

Everything® **Creative Writing Book**
$12.95, 1-58062-647-5

Everything® **Crossword and Puzzle Book**
$12.95, 1-55850-764-7

Everything® **Dating Book**
$12.95, 1-58062-185-6

Everything® **Dessert Cookbook**
$12.95, 1-55850-717-5

Everything® **Diabetes Cookbook**
$14.95, 1-58062-691-2

Everything® **Dieting Book**
$14.95, 1-58062-663-7

Everything® **Digital Photography Book**
$12.95, 1-58062-574-6

Everything® **Dog Book**
$12.95, 1-58062-144-9

Everything® **Dog Training and Tricks Book**
$14.95, 1-58062-666-1

Everything® **Dreams Book**
$12.95, 1-55850-806-6

Everything® **Etiquette Book**
$12.95, 1-55850-807-4

Everything® **Fairy Tales Book**
$12.95, 1-58062-546-0

Everything® **Family Tree Book**
$12.95, 1-55850-763-9

Everything® **Feng Shui Book**
$12.95, 1-58062-587-8

Everything® **Fly-Fishing Book**
$12.95, 1-58062-148-1

Everything® **Games Book**
$12.95, 1-55850-643-8

Everything® **Get-A-Job Book**
$12.95, 1-58062-223-2

Everything® **Get Out of Debt Book**
$12.95, 1-58062-588-6

Everything® **Get Published Book**
$12.95, 1-58062-315-8

Everything® **Get Ready for Baby Book**
$12.95, 1-55850-844-9

Everything® **Get Rich Book**
$12.95, 1-58062-670-X

Everything® **Ghost Book**
$12.95, 1-58062-533-9

Everything® **Golf Book**
$12.95, 1-55850-814-7

Everything® **Grammar and Style Book**
$12.95, 1-58062-573-8

Everything® **Great Thinkers Book**
$14.95, 1-58062-662-9

Everything® **Travel Guide to The Disneyland Resort®, California Adventure®, Universal Studios®, and Anaheim**
$14.95, 1-58062-742-0

Everything® **Guide to Las Vegas**
$12.95, 1-58062-438-3

Everything® **Guide to New England**
$12.95, 1-58062-589-4

Everything® **Guide to New York City**
$12.95, 1-58062-314-X

Everything® **Travel Guide to Walt Disney World®, Universal Studios®, and Greater Orlando, 3rd Edition**
$14.95, 1-58062-743-9

Everything® **Guide to Washington D.C.**
$12.95, 1-58062-313-1

Everything® **Guide to Writing Children's Books**
$14.95, 1-58062-785-4

Everything® **Guitar Book**
$12.95, 1-58062-555-X

Everything® **Herbal Remedies Book**
$12.95, 1-58062-331-X

Everything® **Home-Based Business Book**
$12.95, 1-58062-364-6

Everything® **Homebuying Book**
$12.95, 1-58062-074-4

Everything® **Homeselling Book**
$12.95, 1-58062-304-2

Everything® **Horse Book**
$12.95, 1-58062-564-9

Everything® **Hot Careers Book**
$12.95, 1-58062-486-3

Everything® **Hypnosis Book**
$14.95, 1-58062-737-4

Everything® **Internet Book**
$12.95, 1-58062-073-6

Everything® **Investing Book**
$12.95, 1-58062-149-X

Everything® **Jewish Wedding Book**
$12.95, 1-55850-801-5

Everything® **Judaism Book**
$14.95, 1-58062-728-5

Everything® **Job Interview Book**
$12.95, 1-58062-493-6

Everything® **Knitting Book**
$14.95, 1-58062-727-7

Everything® **Lawn Care Book**
$12.95, 1-58062-487-1

Everything® **Leadership Book**
$12.95, 1-58062-513-4

Everything® **Learning French Book**
$12.95, 1-58062-649-1

Everything® **Learning Italian Book**
$14.95, 1-58062-724-2

Everything® **Learning Spanish Book**
$12.95, 1-58062-575-4

Everything® **Low-Carb Cookbook**
$14.95, 1-58062-784-6

Everything® **Low-Fat High-Flavor Cookbook**
$12.95, 1-55850-802-3

Everything® **Magic Book**
$14.95, 1-58062-418-9

Everything® **Managing People Book**
$12.95, 1-58062-577-0

Everything® **Meditation Book**
$14.95, 1-58062-665-3

Everything® **Menopause Book**
$14.95, 1-58062-741-2

Everything® **Microsoft® Word 2000 Book**
$12.95, 1-58062-306-9

Everything® **Money Book**
$12.95, 1-58062-145-7

Everything® **Mother Goose Book**
$12.95, 1-58062-490-1

Everything® **Motorcycle Book**
$12.95, 1-58062-554-1

Everything® **Mutual Funds Book**
$12.95, 1-58062-419-7

Everything® **Network Marketing Book**
$14.95, 1-58062-736-6

Everything® **Numerology Book**
$14.95, 1-58062-700-5

Everything® **One-Pot Cookbook**
$12.95, 1-58062-186-4

Everything® **Online Business Book**
$12.95, 1-58062-320-4

Everything® **Online Genealogy Book**
$12.95, 1-58062-402-2

Everything® **Online Investing Book**
$12.95, 1-58062-338-7

Everything® **Online Job Search Book**
$12.95, 1-58062-365-4

Everything® **Organize Your Home Book**
$12.95, 1-58062-617-3

Everything® **Pasta Book**
$12.95, 1-55850-719-1

Everything® **Philosophy Book**
$12.95, 1-58062-644-0

Everything® **Pilates Book**
$14.95, 1-58062-738-2

Everything® **Playing Piano and Keyboards Book**
$12.95, 1-58062-651-3

Everything® **Potty Training Book**
$14.95, 1-58062-740-4

Everything® **Pregnancy Book**
$12.95, 1-58062-146-5

Everything® **Pregnancy Organizer**
$15.00, 1-58062-336-0

Everything® **Project Management Book**
$12.95, 1-58062-583-5

Everything® **Puppy Book**
$12.95, 1-58062-576-2

Everything® **Quick Meals Cookbook**
$14.95, 1-58062-488-X

Everything® **Resume Book**
$12.95, 1-58062-311-5

Everything® **Romance Book**
$12.95, 1-58062-566-5

Everything® **Running Book**
$12.95, 1-58062-618-1

Everything® **Sailing Book, 2nd Ed.**
$12.95, 1-58062-671-8

Everything® **Saints Book**
$12.95, 1-58062-534-7

Everything® **Scrapbooking Book**
$14.95, 1-58062-729-3

Everything® **Selling Book**
$12.95, 1-58062-319-0

Everything® **Shakespeare Book**
$12.95, 1-58062-591-6

Everything® **Slow Cooker Cookbook**
$14.95, 1-58062-667-X

Everything® **Soup Cookbook**
$14.95, 1-58062-556-8

Everything® **Spells and Charms Book**
$12.95, 1-58062-532-0

Everything® **Start Your Own Business Book**
$12.95, 1-58062-650-5

Everything® **Stress Management Book**
$14.95, 1-58062-578-9

Everything® **Study Book**
$12.95, 1-55850-615-2

Everything® **T'ai Chi and QiGong Book**
$12.95, 1-58062-646-7

Everything® **Tall Tales, Legends, and Other Outrageous Lies Book**
$12.95, 1-58062-514-2

Everything® **Tarot Book**
$12.95, 1-58062-191-0

Everything® **Thai Cookbook**
$14.95, 1-58062-733-1

Everything® **Time Management Book**
$12.95, 1-58062-492-8

Everything® **Toasts Book**
$12.95, 1-58062-189-9

Everything® **Toddler Book**
$12.95, 1-58062-592-4

Everything® **Total Fitness Book**
$12.95, 1-58062-318-2

Everything® **Trivia Book**
$12.95, 1-58062-143-0

Everything® **Tropical Fish Book**
$12.95, 1-58062-343-3

Everything® **Vegetarian Cookbook**
$12.95, 1-58062-640-8

Everything® **Vitamins, Minerals, and Nutritional Supplements Book**
$12.95, 1-58062-496-0

Everything® **Weather Book**
$14.95, 1-58062-668-8

Everything® **Wedding Book, 2nd Ed.**
$14.95, 1-58062-190-2

Everything® **Wedding Checklist**
$7.95, 1-58062-456-1

Everything® **Wedding Etiquette Book**
$7.95, 1-58062-454-5

Everything® **Wedding Organizer**
$15.00, 1-55850-828-7

Everything® **Wedding Shower Book**
$7.95, 1-58062-188-0

Everything® **Wedding Vows Book**
$7.95, 1-58062-455-3

Everything® **Weddings on a Budget Book**
$9.95, 1-58062-782-X

Everything® **Weight Training Book**
$12.95, 1-58062-593-2

Everything® **Wicca and Witchcraft Book**
$14.95, 1-58062-725-0

Everything® **Wine Book**
$12.95, 1-55850-808-2

Everything® **World War II Book**
$12.95, 1-58062-572-X

Everything® **World's Religions Book**
$12.95, 1-58062-648-3

Everything® **Yoga Book**
$12.95, 1-58062-594-0

*Prices subject to change without notice.

EVERYTHING SERIES!

Everything® **Kids' Baseball Book, 2nd Ed.**
$6.95, 1-58062-688-2

Everything® **Kids' Cookbook**
$6.95, 1-58062-658-0

Everything® **Kids' Joke Book**
$6.95, 1-58062-686-6

Everything® **Kids' Mazes Book**
$6.95, 1-58062-558-4

Everything® **Kids' Money Book**
$6.95, 1-58062-685-8

Everything® **Kids' Monsters Book**
$6.95, 1-58062-657-2

Everything® **Kids' Nature Book**
$6.95, 1-58062-684-X

Everything® **Kids' Puzzle Book**
$6.95, 1-58062-687-4

Everything® **Kids' Science Experiments Book**
$6.95, 1-58062-557-6

Everything® **Kids' Soccer Book**
$6.95, 1-58062-642-4

Everything® **Travel Activity Book**
$6.95, 1-58062-641-6

Available wherever books are sold!
To order, call 800-872-5627, or visit us at everything.com

Everything® is a registered trademark of Adams Media Corporation.